STORIES FROM SCIENCE

Stories from Science

BY A. SUTCLIFFE & A. P. D. SUTCLIFFE

WITH ILLUSTRATIONS BY
ROBERT HUNT

Book 2

CAMBRIDGE: AT THE UNIVERSITY PRESS

PUBLISHED BY
THE SYNDICS OF THE CAMBRIDGE UNIVERSITY PRESS
Bentley House, 200 Euston Road, London, N.W.1
American Branch: 32 East 57th Street, New York 22, N.Y.
West African Office: P.O. Box 33, Ibadan, Nigeria

©

CAMBRIDGE UNIVERSITY PRESS
1962

Printed in Great Britain at
The Stellar Press, Barnet, Herts.

CONTENTS

PREFACE

One of the authors, when a young science master in Cambridge, decided to collect stories of unusual incidents or chance occurrences in science and engineering which would enrich his teaching and interest his pupils. So began a hobby of collecting stories which has given him much pleasure during the last forty years. With the help of his son, these stories have been prepared for publication in the hope that they will give similar pleasure to others.

It will be obvious that in order to gather all this information together, reference has been made to a variety of sources. A list of some of the books and articles consulted is given at the end of each volume, and the authors desire to express their sincere thanks to all those writers whose works have proved so helpful.

The illustrations, which add so much interest to the book, are the work of Mr Robert Hunt, who has successfully combined great accuracy of detail with artistic skill.

The authors gladly acknowledge their gratitude to many colleagues and friends: to Mr G. H. Franklin who translated numerous passages, and to Mr L. R. Middleton, Mr J. Harrod, Dr A. H. Briggs, Dr R. D. Haigh and Miss M. Lipman, who read the typescript.

They are particularly indebted to Mr R. A. Jahn for that severe yet constructive criticism which only a friendly colleague of many years' standing can give. In its final stages before printing the book owes a great deal to the officers of the University Press for their very helpful suggestions and emendations.

A. S.

A. P. D. S.

LINCOLN 1962

7

23. Archimedes: The Scientific Detective

ARCHIMEDES was born in the year 287 B.C. at Syracuse, the most important city of ancient Sicily, where the Greeks had founded a settlement some five centuries before his birth. He followed the custom of gifted youths in ancient Greece of studying at the royal school of Alexandria, in Egypt, where the best education of his day was given in mathematics and physics. On his return his talent for applying his theoretical knowledge to practical problems gained him the favour or patronage of the King of Syracuse, to whom he was probably related.

Hiero II of Syracuse was a valiant warrior and a devout worshipper of the Gods, and he celebrated most of his victories on the battlefield by making offerings to one or other of his Gods. For example, after one victory he built a temple and after another erected a public altar.

To celebrate one of his triumphs he decided to present a very valuable gold crown to the Temple of the Immortal Gods. An expert goldsmith was appointed to make the gift and the king's treasurer gave him a certain weight of the precious metal. At the appointed time the goldsmith delivered the crown for the King's inspection. The King was well satisfied.

Before long, however, as the result of information received, Hiero became suspicious that the goldsmith had not used all the gold to make the crown but had kept some, replacing it by silver. But he did not know how to detect whether this had been done or not. The crown most certainly weighed the same as the gold delivered to the goldsmith. Little information was to be gained simply by taking a look at it, because when gold and a little silver are melted together the solid which forms is yellow in colour and not unlike pure gold in appearance.

Hiero did not wish to spoil the beautifully made crown merely to decide whether a rumour was true or false. So he decided to call in Archimedes to investigate the alleged crime.

9

Archimedes carefully considered the problem but at first could think of no way of solving it. One day, however, while the problem was very much in his mind, he happened to go to the public baths. The tub was full of water, and he observed that water flowed out as his body entered the tub.

Archimedes in his bath

Countless people before him had also noticed this; and no doubt Archimedes himself must have seen the water overflow on numerous occasions without paying any attention to it. But on this day his mind was set on a method of solving the king's problem, and in an instant he saw how to do it. He reasoned that the volume of the water which was pushed out of the tub was the same as the volume of his body which entered it. Hence, if he submerged the golden crown in a vessel brimful of water a volume of water equal to the volume of the crown would be forced out of the vessel.

Archimedes was so excited by his new discovery that he did not wait to take his bath, but jumped out of the tub. He forgot that he was naked and ran home through the streets, shouting

23. *Archimedes: The Scientific Detective*

many times with glee, '*Eureka! Eureka!*', a Greek word which means '*I have found it*'.[1]

Soon he was at work on his new idea. He already knew, as indeed did many others, that gold is one of the densest of all metals, a piece of it being much heavier than a piece of silver of the same size. He knew also how to measure accurately the volume of a piece of gold or silver of a regular shape, like a cube or brick, by measuring its length, breadth and height and multiplying the three measurements together. But the difficulty before he hit on his brilliant idea had been how to measure the volume of an irregular shape such as a crown.

His new method was very simple. First of all he weighed the crown carefully, then he obtained a lump of pure gold and a lump of pure silver, each weighing exactly the same as the crown. He next filled a vessel to the brim with water and carefully lowered the lump of gold into it. Some of the water overflowed and he measured how much had done so. This volume of water, he reasoned, was the same as the volume of the piece of gold.

Archimedes repeated this experiment, using the lump of silver instead of that of gold. He found, as he had expected, that the volume of water which had overflowed this time was greater than when the gold had been used.

He then lowered the crown into the full vessel and found its volume in the same way. This volume was greater than that of the gold but less than that of the silver. It was obvious to Archimedes that the crown was not pure gold, as it should have been. He even had enough information to calculate how much gold had been replaced by silver.

The dishonest goldsmith had been found out. What dire punishment was meted out to him does not seem to have been recorded but it is said that when he heard how Archimedes had detected his crime he at once confessed his guilt.

*   *   *   *   *

Today, Archimedes' joyful run through the streets of Syracuse is still commemorated by the name *eureka tin* which is given to a vessel with a spout used in the laboratory. The tin is filled with water as far as the spout. The solid whose volume is to be determined is carefully lowered into it. The volume of water which is displaced runs down the spout and into a measuring cylinder placed below it. This simple way of finding the volume of irregularly shaped solids, such as pieces of stone, is still used in many laboratories.

24. Archimedes: The Military Engineer

IN ANCIENT times the port of Syracuse was a flourishing and important city with its own king and armed forces. As mentioned in chapter 24, it is situated on the island of Sicily, and since it is not far from Rome it could serve as a good base for an enemy of Rome – such as Carthage, a large city on the North Coast of Africa (chapter 1).

In the year 214 B.C. the king of Syracuse allied himself with the Carthaginians; so, to prevent the latter from using Syracuse as a base, the Romans sent their best general, Marcellus, to seize it. Hiero II, King of Syracuse, had foreseen such an attack and had taken precautions to fortify the city by appointing as his chief military engineer his friend and kinsman, Archimedes.

Archimedes was chosen for this post because he had made a careful study of mechanics. He had designed the lever (or crowbar), the system of pulleys and many other machines. To emphasise the great power which could be exerted by the crowbar, he had once said, 'Give me another place on which to stand and a lever long enough and I will move the earth'.[1] When the king heard of these machines he commanded Archimedes to show him what they could do. Archimedes chose for his demonstration one of his pulley systems and a three-masted ship.

He connected a long rope to his pulley system, tied one end of it to the ship and whilst holding the other end walked away from the ship. He then sat down on the sands and while the spectators watched pulled gently on the rope. The ship, we are told, 'glided towards him with as smooth and even a motion as if it were sailing a calm sea'.[2]

All were amazed, for they had never seen a pulley block at work before and it seemed almost a miracle that one man could do the work of many others so easily. The king at once realised the value of Archimedes' knowledge and asked him to construct a number of war machines, some for attack and some for the defence of a city. This Archimedes did, although it is said that he did not consider the construction of any of these engines as serious labour, but as 'the mere holiday sport of a geometrician'.

Syracuse is situated on a peninsular with a long sea coast, and Marcellus, the Roman general, launched his attack by land and sea at the same time. Unhappily for him, however, 'he had not made due reflection upon the great skill of Archimedes, nor considered that the mind of a single man is on some occasions superior to the force of many hands. But this truth was soon discovered'.[3]

The soldiers of Syracuse had been well trained in the use of the war-engines. They showered on the attackers darts of all kinds, as well as large stones, striking them down in heaps and throwing their ranks into confusion. Some of the machines went off with such a bang that the enemy were greatly frightened, and a legend has since grown that Archimedes had discovered gunpowder and was using it. In all probability the noise was made by the powerful springs and levers which were brought into action when the machines hurled a large stone. The engines were so effective that the land attack had to be called off.

The attackers by sea had an equally warm reception. Archimedes had invented a machine which was composed of a long heavy wooden pole, hanging from supports at each end, like the long seat of a swing. The machine was stationed near one of the gates in the sea wall and when a ship approached this spot the soldiers rushed to the machine, pulled and pushed the pole to and fro until it was swinging strongly and then quickly opened the gate and directed the swinging pole against the sides of the enemy's ship, which it smashed to pieces.

Another machine consisted of a long beam which rested on a pivot placed on top of the walls, and was balanced like a see-saw, with half of it projecting over the sea-wall. Ropes were fastened to the city end of the beam and a large iron claw to the other end. When a suitable opportunity presented itself the soldiers inside the walls pushed up their end of the beam so that the other end was lowered. They carefully manœuvred the beam until the iron claw caught hold of one of the enemy's ships. Then they quickly pulled on the ropes and so lifted the ship high out of the water. They then released the claw. An ancient writer des-scribed the scene thus:

Often was seen the fearful sight of a ship lifted out of the sea into the air, swaying and balancing about until the men were all thrown out of it or overwhelmed with stones fired at them from slings. The empty vessel was then either dashed against the walls or released to drop from a great height into the sea.

To force an entry over the sea walls, Marcellus had relied on a machine called the sambuca. This consisted of a long ladder, with a platform at the top. It was carried on a stage resting on a number of small ships so that it could be sailed close to the walls of a besieged city. In many ways it was like the fireman's ladder of today. It was placed almost upright near the walls, so that it was nearly touching them, and then a few men climbed up the ladder to the platform at the top. On this platform there was a small landing stage which could be pushed out until it rested on the walls of the city. When his had been done, other soldiers from the ships climbed up the ladder, ran across the landing stage and jumped into the city.

Archimedes knew all about this sambuca and 'held his fire' until the sixteen ships which carried it were within range of a gigantic catapult which he had made and which was capable of hurling a stone weighting ten talents (half a ton, according to some estimates). When the sambuca was near enough the men let off the catapult. The stone landed with a terrible crash, breaking the stage on which the sambuca rested and making huge holes in the ships carrying it. No wonder the attack by sea fared no better than that by land.

The gigantic catapult

Marcellus renewed the sea attack before daybreak, hoping to get his soldiers to the foot of the walls unnoticed. But Archimedes had expected this ruse and his men purposely kept quiet until the Romans had reached their objective and a large number of them had gathered below the walls. Then showers of missiles were hurled down from his new machines on to their heads, inflicting great losses. The Romans retired in disorder; many declared that they were fighting against gods and not men, since destruction fell upon them from invisible hands.

Marcellus tried to instil courage into his soldiers, saying, 'Are we to give in to this geometrician who sits at ease by the sea shore and plays at upsetting our ships, to our lasting disgrace, and surpasses the hundred-handed giant of the fairy story by hurling so many weapons at us at once?'[4] But the common soldiers had become so scared that when they saw even a rope or beam of wood being put out over the wall top they turned and fled, crying out that Archimedes was bringing up another new-fangled machine.

Meantime, according to L. J. Tzetzes, a twelfth-century writer, another 'machine' invented by Archimedes was at work on the ships as they stood a little distance out to sea. This machine was made of a large number of mirrors mounted on wood. It will be recalled that when the rays of the sun fall on a mirror they are thrown back or reflected. Archimedes, it is thought, used a large plain mirror with numerous small mirrors fixed on its sides, and hinged so that they could be moved as desired. The large mirror was placed to catch and reflect the sun's rays on to one of the wooden ships of the enemy. Then each of the small mirrors was moved until it too reflected the sun's rays on the same spot. The heat given off by all these mirrors, concentrated on one spot, was sufficient to set fire to any wooden ship within a 'bow shot's distance from the walls'.

All these new-fangled war machines served their inventor's purpose well and the first assaults on the city failed. Marcellus withdrew his attacking forces but he did not give up the fight. Instead of attempting a direct attack on such a well-defended city, he blockaded it by preventing supplies of goods of all kinds from entering or leaving. Finally, after a siege of some three years' duration, he decided to make another effort to capture the town. Even then he did not plan to attack it directly, for he still feared Archimedes' machines. Instead he decided to try treachery.

He succeeded in getting a few of the citizens to help him, and one night these traitors inside the city walls admitted some of the

Roman soldiers. The men of Syracuse had become lax in their guard and thus, in a short but savage attack, the city was captured. As was customary in those days, the victorious soldiers were given permission by the leaders to plunder the city, but Marcellus especially ordered them to spare the lives of the important citizens. Notwithstanding this order, his soldiers put many prominent inhabitants to the sword and unfortunately amongst those killed was Archimedes.

Different versions of his death are given.[5] One of these states that he had drawn a geometrical figure on the sands of the seashore and was studying it so intently that he did not notice the assault of the Romans and the capture of the city. A soldier suddenly appeared before him and ordered him to present himself to Marcellus, but he refused to do so until he had solved the problem. Thereupon the man, in a rage, drew his sword and killed him.

The death of Archimedes

Others say that this Roman soldier fell upon Archimedes with a sword, intent on killing him. Whereupon Archimedes begged him to wait for a little while so that he might not leave his theorem imperfect. But the soldier refused to do so and killed him at once.

Another version is that Archimedes was carrying boxes containing his mathematical instruments, including sundials and quadrants, when some soldiers met him, and, supposing that there was gold in the boxes, slew him in order to obtain it.

Whatever the manner in which it was brought about, all accounts agree that Marcellus was much grieved when he heard of Archimedes' death.[6]

$$*\qquad *\qquad *\qquad *\qquad *$$

These are some of the legends and stories about the wonderful machines which Archimedes is said to have invented. Other evidence shows that some of the machines mentioned were in use long before Archimedes' time. Thus it is recorded that Philip II of Macedonia, who lived from 382 B.C. to 336 B.C., used machines such as battering rams and a catapult for hurling heavy stones. Archimedes' inventions, however, are mentioned by many writers, including some who lived about the same time, and all of them give more or less the same descriptions; thus it seems likely that he did invent many of the machines.

The use of 'burning glasses' to set fire to articles was also known long before the days of Archimedes. For example, mention of it was made in *The Clouds*, by Aristophanes, which was written two hundred years before the siege. In this comedy an actor tells another about a debt he owes, saying that if a note is made on the tablet about it he will get a burning glass and burn out every line of the writing. In those days a tablet was made of wax and the words were scratched on it. The man could therefore have used a burning glass to melt the surface of the wax and thus erase the writing.[7]

In the year 1727, Buffon, the French naturalist, reconstructed the kind of apparatus which Archimedes might have used. To the sides of a large, plain, hexagonal mirror he hinged 168 small mirrors and placed the apparatus in the path of the sun's rays. He then moved each mirror so that all the reflected rays were focused on a spot 150 feet away where he had previously placed a pile of tinder. The heat set the tinder on fire. He repeated the experiment but focused the spot on to some lead 140 feet away. The lead melted.[8]

Many years before these experiments were made another philosopher named Kircher had done similar experiments and had also visited Syracuse. On viewing the harbour he came to the conclusion that the galleys of Marcellus could not have been more than thirty paces away from the walls and so would be well within 'range' of the mirrors' focusing point. Indeed Plutarch recorded that some of the enemy ships sailed near enough to the walls of the city for the defenders to use grappling irons on them; so it would seem that they would also be near enough for the burning glass to have some effect.

The eminent mathematician, Rouse Ball, in commenting on Buffon's experiments, pointed out that they were made in Paris and in the month of April, and concluded that 'in a Sicilian summer the use of several mirrors might be a serious annoyance to a blockading fleet, if the ships were sufficiently near'.[9]

It must be pointed out, in conclusion, that Buffon's experiments only show that the method could have been successful had it been used; they do not prove that Archimedes did, in fact, actually use the burning glass. The story would have been more convincing than it is had it been told by an author living at the time of Archimedes, or by one who came shortly afterwards. No reference was made to a burning glass by Plutarch, Livy or Polybius, although each of them described the mechanical war engines invented by Archimedes.

25. Buried in Mid Air

MOHAMMED, the prophet, was born of Arab parents and spent most of his early life looking after the flocks of sheep and herds of camels just as other Arab boys did in his day.

As he grew older he began to think more and more about God. One day, when he was forty years old, he had a vision, and dreamed that the angel Gabriel was calling him to go 'out into the world' and teach people about the living God. This he did. At first he had but few followers but before his death they numbered thousands upon thousands and became known as Moslems. They included Arabs living in Mesopotamia, who were called Saracens, as well as people living as far apart as India and Northern Africa.

Mohammed preached that there was only one God: a God who was a loving father to those who believed in him, but a cruel tyrant to those who did not. His belief was expressed simply by the saying, 'There is no God but Allah, and Mohammed is his prophet'; and he ordered his followers to kill without mercy all unbelievers who would not change their faith when commanded to do so.

Many legends have grown up about such a famous person and the following one, told by an Italian writer of the fifteenth century, was commonly believed for centuries.

When Mohammed died his body was carried by the Saracens into a city of Persia and put into a coffin of iron, which hung unsupported in the air. It was in fact suspended there by the attraction of a loadstone, but those who did not know the properties of that stone believed a miracle had taken place.[1]

The property of loadstone referred to is its attraction for iron. Loadstone is a rock consisting mainly of the black oxide of iron. It occurs in a few countries and occasionally small pieces of it are found exposed as outcrops on the surface of the ground.

Pliny tells a story of the discovery of its magnetic properties by a shepherd named Magnes.[2] One day while walking behind his sheep on the slopes of Mount Ida, in Asia Minor, Magnes happened to put his foot on an exposed piece of black rock. To his surprise the iron nails of his shoes and the iron ferrule of his staff stuck fast to it. For this reason the stone became known as Magnes stone.

There are a few other stories of a similar kind but all of them are now regarded as legendary. Yet it is very probable that the

25. Buried in Mid Air

magnetic property of this stone was discovered in some such accidental manner. One of these stories places the discovery in the hills of the ancient country of Magnesia, a part of Asia Minor; – hence, according to this story, the name magnetite.

Other early references to the stone mention its use as a compass, for when a small piece of it is suspended in the air and left to come to rest, it points in a north-south direction. Travellers of centuries ago used it to show the way; so in England it became called loadstone, 'load' coming from an old English word meaning 'way'.

It is said that this property of loadstone was known in China three thousand years before the birth of Christ and that Chinese sailors then used the stone as a navigational aid.

For centuries after Mohammed's death many Christians believed that the Moslem architects who had planned his tomb had made use of the loadstone's property of attracting iron by inserting pieces of the stone in the roof and in the floor of the vault.[3] This insertion, they believed, had been done so skilfully that the iron coffin remained at rest, suspended in air, between the roof and the floor of the vault.

There was no easy way for Christians to check this belief because the Moslems dealt summarily with any of them who visited their lands. They kept a close watch for 'unbelievers' (who were called 'infidels') in order to convert them to their religion. Any such person who was captured was given the option of becoming a Moslem or being put to death. If the person agreed to change his faith he was, nevertheless, compelled to live in a Moslem country and carefully watched so that he could not return to his native land and become a Christian once more. Hence very few Christians who visited Medina, the city where Mohammed was buried, ever got back to Europe.

In 1513, however, an Italian who had made good his escape, described Medina and the prophet's tomb. He stated that he had seen the 'ark or tomb of wicked Mohammed' and that the coffin did not hang in the air.[4]

Many years later an English youth, 'Jose Pitts of Exon', was captured by pirates, made a slave and compelled to be a Moslem.

After many years of captivity he too escaped and wrote an account of Medina, part of which ran thus: 'It is stated by some that the coffin of Mohammed hangs up because of the attraction of loadstone to the roof of the mosque; believe me it is a false story. When I looked through the brass gate I saw the tops of the curtains covering the tomb.' He then stated that these curtains did not reach half the way to the roof and that he could see nothing hanging in the space between them and the roof.

But the story was still believed by most people even when, in 1737, a well-informed writer said that 'this romantic story makes Moslems laugh heartily when they know that Christians tell it as a certain fact'.

There is now no doubt about Mohammed's tomb, the following account being generally accepted as giving the facts. Shortly before he died, Mohammed

expressed the opinion that any prophet should be buried in the place where he died. This was complied with to the very letter, for a grave was 'digged' in the house of Ayesha (his wife), beneath the very bed on which he had died.

It has since been included in a spacious temple, with an enclosure, surrounded by an iron railing, painted green, wrought with filigree work and interwoven with brass and gilded wire, admitting no view of the interior excepting through small windows about six inches square. Above this sacred enclosure rises a lofty dome surmounted with a gilded globe and crescent, at the first sight of which, pilgrims as they approach Medina salute the tomb of the prophet with profound inclinations of the body and appropriate prayers.'[5]

The idea of suspending an iron object in the air by using loadstones is a very old one. Indeed it is recorded that one of the kings of Ancient Egypt ordered his architect to make an iron statue of his dead sister and then to hang it in mid air under the ceiling of a vault 'crowned with loadstone'. But both the king and the architect died before the attempt was made.

Another story tells that

an image of the sun was artfully made of iron; a piece of loadstone was then placed in the roof of the Temple and the iron sun was thus made to be hanging seemingly in mid air without any support; but a worthy servant of God, having discovered the trick, took the loadstone out of the roof; upon which the iron sun immediately dropped to the ground and was broken into a thousand pieces.[6]

25. Buried in Mid Air

There is no doubt, therefore, that many people even before the birth of Christ believed that iron could be suspended in mid-air by the use of loadstone.

Early in the seventeenth century two writers considered how this could be done. One of them wrote that 'nothing can so hang in the air by the force of the loadstone' unless it 'either touches the stone itself or some other intermediate substance between it and the stone'. For example, he continued, 'lay two or three needles upon a smooth table, put a silver or pewter plate upon them and upon the plate a loadstone'. Then when the plate was lifted a few inches from the table it would be seen that the needles were hanging in the air but the tops of them were touching the under surface of the plate.

'A multitude of magnets', he wrote, 'such as would be needed to support a heavy weight would but confound each other's force.' They would act 'much like a team of horses where every one drawing his sundry way might soon with disordered stretching tire himself and his fellows but never move the load one iota from the place'.[7]

The other man to consider this problem was Father Cabeus, (1585–1650), who used experimental methods in trying to solve a problem. He thought of a very neat but delicate experiment. An old account of it reads as follows: 'He placed two loadstones above one another and distant about four fingers; then taking a needle by the middle with his two fingers, he gently carried it between the two loadstones, endeavouring to find the spot where the needle, not being attracted by the one loadstone more than the other would remain suspended in the air without being supported.' After trying again and again Father Cabeus at last succeeded in placing the needle at the ideal spot. 'The needle continued in the air between the two loadstones without touching anything and this surprising sight continued as long as one might repeat four long verses.' But when he arose to call some of his friends, 'the motion of the air broke, as one may call it, the charm'.[8]

Father Cabeus himself stated that he was successful in suspending the needle. But even if we can accept this statement we

The needle rests in mid air

cannot assume from his experiment that a heavy rectangular iron coffin could be suspended in a similar way between more powerful loadstones. Moreover, whilst it is possible to obtain loadstones powerful enough to attract a needle and support its small weight, it seems almost impossible to find loadstones which can support a weight of many hundredweights. The most powerful loadstone known to science was probably that presented to King John of Portugal by an Emperor of China which was reputed to be capable of supporting a weight of three hundred pounds.[9] Such a loadstone is very rare and quite a few of them would be needed to support a heavy iron coffin.

The effect of gravity on the coffin would make the exact placing of the loadstones very difficult. For all the forces on the coffin would have to be in equilibrium if the coffin were to rest horizontally and it would have looked unseemly if it had been tilted so that one end or one side was higher than the other.

It is most unlikely that any architect could design the building of a dome and a floor studded with loadstone that would fulfil these conditions. Hence the following explanation, given by one of the best-known writers on Mohammed, may well be the true one: 'In order to keep the coffin clear of the floor of the vault it was supported on nine bricks, the earth being heaped about its sides. That is the entire extent to which the coffin was suspended in air, namely by nine bricks being put under it.'

This story of Mohammed's tomb is matched by other legendary stories about the power of loadstone to attract iron. A popular one which circulated for centuries is based on a belief in the existence of black rocks whose magnetic power was strong enough to pull the nails out of ships passing close to them. Typical of these is the story by the Arab author of *The Thousand and One Nights*, an abridged account of which runs as follows:

I was a King, and took pleasure in sea voyages, and therefore embarked with a fleet of ten ships. I proceeded twenty days, after which there arose against us a contrary wind. We found ourselves in strange waters, unknown to the captain, and looking towards the midst of the sea I perceived something looming in the distance, sometimes black, and sometimes white.

When the captain heard of this, he threw his turban on the deck, and plucked his beard, and said to those who were with him: 'Receive warning of our destruction, which will befall all of us; not one will escape. O my lord, know that we have wandered from our course; tomorrow we shall arrive at a mountain of black stone, called loadstone; the current is now bearing us violently towards it, and the ships will fall in pieces, and every nail in them will fly to the mountain, and adhere to it; for God hath given to loadstone a secret property by virtue of which everything of iron is attracted towards it. On that mountain is such a quantity of iron as no one knoweth but God, whose name be exalted; for from times of old great numbers of ships have been destroyed by the influence of that mountain.'

On the following morning we drew near to the mountain; the current carried us towards it with violence, and when the ships were almost close to it, they fell asunder, and all the nails, and everything else that was of iron, flew from them towards the loadstone. It was near the close of day when the ships fell in pieces. Some of us were drowned, and some escaped, but the greater number were drowned.[10]

Several other Arab writers describe this mountain of loadstone. One says it is on the shore of the Indian Ocean and that if the ships which navigate this sea and contain anything of iron approach the mountain, 'it flies from them like a bird and adheres to the mountain; for which reason, it is the general custom to make use of no iron in the construction of the vessels employed in this navigation'.

25

The black mountain was located by various other writers in places as far apart as the Indian Ocean, the Mediterranean and even Greenland. The myth survived until as late as the sixteenth century.

26. The Unexpected Behaviour of the Compass Needle

THE HISTORY of science records two noteworthy occasions when the needle of a compass has behaved in a most unexpected manner. The first of these took place at sea, in the year 1492, during the voyage of Columbus in search of the Indies. The second took place inside a lecture room in 1819, when a university professor was delivering a lecture to undergraduates.

*   *   *   *   *

Columbus, like other mariners of his time, relied on the heavenly bodies and the magnetic compass to guide him over the seas when out of sight of land. He knew that the Pole Star was in approximately the same position each night, and used it as a guide. He also knew that while the compass needle pointed approximately north and south it did not point *exactly* to the Pole Star.

Columbus did not sail immediately into the unknown when he set out on Friday, 3 August 1492. For he took a course which led to the Canary Islands, a voyage which a few other sea captains had made before him. He stayed there for three weeks and then, on 6 September, he left and steered westwards into the vast ocean of unexplored water. One account of what then happened reads:

Three days later all land had passed from sight and, on losing sight of this last trace of land, the hearts of the crews failed them. They seemed literally to have taken leave of the world. Behind them was everything dear to the

26

26. *The Unexpected Behaviour of the Compass Needle*

heart of man: country, family, friends, life itself; before them everything was chaos, mystery and peril. In the perturbation of the moment, they despaired of ever more seeing their homes. Many of the rugged seamen shed tears, and some broke into loud lamentations. The admiral tried in every way to sooth their distress, and to inspire them with his own glorious anticipations. He described to them the magnificent countries to which he was about to conduct them: the islands of the Indian seas teeming with gold and precious stones: the regions of Mangi and Cathay, with their cities of unrivalled wealth and splendour. He promised them land and riches and everything that could arouse their cupidity, or inflame their imaginations, nor were these promises made for purposes of mere deception, he certainly believed that he should realise them all.[1]

Then, about a week after they had left the Canary Islands, Columbus noticed that the compass needle was not pointing in the direction he had expected. Next morning it was even further out of its customary direction. He was very much surprised and

Columbus and the Compass

puzzled, and more so when on each of the next three days the variation from the normal direction increased. He mentioned it to nobody for he knew how depressed all his men were and did not wish to alarm them further. But he knew that he could not keep the secret for long because one or the other of his pilots

27

would be certain to notice the compass's strange behaviour before many days had passed. As he expected, one of them did notice what had happened, and as soon as the men heard of it they were filled with a great fear. It seemed to them that the compass needle was losing its power in this unknown world just at the time when they most needed its help. They knew that without a compass to guide them they would soon get lost in the vast and trackless ocean. They feared that if a compass, normally so reliable, began to behave in so unusual a way in the new world they were entering then everything else might do so as well.

Columbus, however, had got his story ready by that time and had realised that he must not 'blame' the compass needle. So he declared that the compass needle still retained its power but that the Pole Star, in this part of the world, changed its position. He told them that it moved in a circle about the true north. Columbus had previously earned for himself a very great reputation as an astronomer and the men believed his explanation. Confidence was thus restored and the men were no longer afraid.[2]

Another account of the incident, written by the Spanish historian Oviedo, gives more information about the behaviour of the crew, stating that

they were so incensed and frightened by the behaviour of the needle that they wished to throw Columbus overboard and that they greatly resented the action of Ferdinand and Isabella of Spain in entrusting them to the care of such a man. They became rebellious and cried repeatedly 'Turn back to Spain'.

As a story is told and retold, time and time again, the original version often becomes very much changed. It is, therefore, interesting to turn to the first printed account of this incident which was written by Columbus's son, Ferdinand, and was based on entries in his father's diary for the year 1492. It reads;

On the 13 of September, he found that at nightfall, the needle varied half a point towards the North East, and at break of day, half a point more, by which he understood that the needle did not point at the North Star, but at some other fixed and invisible point. This variation no man had observed before, and therefore he had occasion to be surprised at it. But he was more amazed the third day after, when he was almost a hundred leagues further,

26. *The Unexpected Behaviour of the Compass Needle*

for at night the needles varied about a point to the North East, and in the morning they pointed upon the star.[3]

Some accounts of the incident mention that Columbus was believed by his men because of his high reputation as a skilled navigator. But, according to A. Crichton Miller, an authoritative writer on Columbus, there is not sufficient reliable information about the early life of Columbus to be sure that he had great skill in navigation. Indeed this writer believes that it is probable that 'with regard to terrestrial magnetism, as then understood, he had no closer acquaintance than the average pilot of his time'. He added: 'If this view is correct, he was not in a position to place an intelligent interpretation on the variations exhibited by compass needles during the crossing of the Atlantic'.

These stories, and similar ones, have led many people to believe that Columbus discovered the variation of the compass, a belief which has been enlarged upon by many writers. It is, however, 'almost certain that an easterly declination had been observed in North Western Europe before Columbus sailed on his first voyage', according to the writer already quoted.[4]

Even so, Columbus was probably the first to give any approximately correct record of the variation of the compass during long sea voyages in easterly or westerly directions.

The story of the next noteworthy occasion of the unexpected behaviour of a compass needle begins seven years after Volta's discovery of a method of obtaining current electricity (chapter 35). The English scientist, Sir Humphry Davy, had used such a current to decompose a chemical substance and thereby obtained a new metal called sodium. His discovery, made in 1807, led scientists to study the chemical effects of the electric current on many substances and so for a time little attention was paid to the other properties of electricity.

In 1819, however, a lucky chance led to the discovery of certain mechanical properties of current electricity which were to be of tremendous value to science and industry.

Professor John Christian Oersted, professor of physics at Copenhagen University, while lecturing on electricity, galvanism and magnetism, was using a long wire connected to the terminals

of a voltaic pile. In the course of this lecture he said, quite casually, 'Let us now, as the battery is in activity, try to place the wire parallel with the needle'. He then put the wire over a compass needle and parallel to it, and switched on the current. To his amazement he saw the needle turn at a right angle to the wire.

Oersted at once realised that this unexpected occurrence was worth a full investigation, and with a friend repeated the experiment and extended it. But because his original experiment was made 'with feeble apparatus' they used a much stronger battery. This he described as follows:

The galvanic apparatus which we employed consisted of twenty copper troughs, the length and height of which was twelve inches, but the breadth scarcely exceeded two and a half inches. Every trough is supplied with two

Oersted's experiment

plates of copper, so bent that they could carry a copper rod, which supports the zinc plate in the water of the next trough. The water of the troughs contained one sixtieth of its weight of sulphuric acid, and an equal quantity of nitric acid. The portion of each zinc plate sunk in the water is a square whose side is about ten inches in length. The opposite ends of the galvanic battery were joined by a metallic wire.[5]

In the picture can be seen the tops of some of the twenty troughs of this voltaic pile; it also shows what cumbersome apparatus scientists then had to use to get electricity. The diagram illustrates the results obtained from his experiments. When the current was flowing in the direction shown by the dotted arrow in (*a*), and the wire was placed above the compass, the needle turned at right angles in the direction shown by the small arrows.

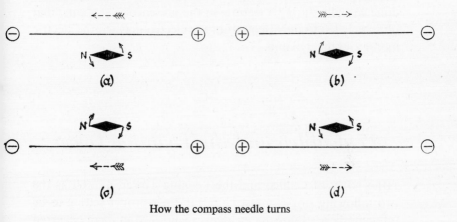

How the compass needle turns

But when the current was reversed as in (*b*) the needle turned in the opposite direction, Oersted then placed the wire below the compass and the needle turned at right angles in a direction depending on the direction of the current, as will be seen by comparing diagrams (*c*) and (*d*).

The publication of his discovery, we are told, 'caused an immense sensation everywhere; it was at once translated and published in scientific reviews of various countries. Not only were his experiments repeated everywhere, but they also inspired a whole series of treatises'.

Soon followed more discoveries by other scientists. It was

shown that the electric current could induce magnetism in iron. Then came the invention of the electro-magnet which was made by winding round a piece of iron a length of insulated wire whose ends were connected to the terminals of a voltaic pile; when the current was switched on the iron became strongly magnetic.

Next followed the discovery that whilst an electric current flowing along a wire produced a magnetic field the reverse also happened: that is, a moving magnet would induce an electric current in a coil of wire.

Faraday (chapter 44) truly said that Oersted's discovery 'burst open the gates of a domain of science, dark till then, and filled it with a flood of light'.[6] In fact it was his experiment done casually during a lecture, in the presence of students, that ultimately led to the discovery of the electro-magnet, the electric motor and the dynamo.

27. Galileo and the Leaning Tower

THE STORY of Galileo and the Leaning Tower, as well as the two following stories, must be put in their proper setting to be fully appreciated, for the incidents mentioned in them occurred at a noteworthy time in the history of science.

Until about the fifteenth century, scholars, with a few exceptions, had accepted without question the teachings of the ancient writers. But in the fifteenth and sixteenth centuries important discoveries were made and changes of many kinds took place. New lands, such as America, were discovered; religion went through a great upheaval with the Reformation; printing was invented; and a few scholars who had the curiosity to examine Nature often obtained surprising results.

About the year 1500 a Polish philosopher named Copernicus startled the intellectual world by his views that the sun was the

27. Galileo and the Leaning Tower

centre of the universe and that the earth moved round it. This view was contrary to the teachings of the ancients. Such a new belief was not very popular nor was it universally accepted, and most of the universities and schools continued to teach the traditional science as laid down by the ancients, and especially by Aristotle, the Greek who lived 350 years before Christ (chapter 29).

Galileo was born in the year 1564. As a young man he studied medicine at first, but early in his university life he decided to change to the study of mathematics. In this subject he showed so much originality that before long his approach to the subject had become quite different from the customary one of merely reading the works of Aristotle and other ancient writers and then discussing them.[1] Galileo could satisfy his curiosity only by making experiments. A few men before him had also studied science by the experimental method and had met with much opposition from the scholars of their day. Galileo did not escape criticism himself, as the following traditional story relates.

In the year 1590 Galileo, who was then a young man of twenty-five and a lecturer in mathematics at the University of Pisa in Tuscany, Italy, decided to make a public experiment on the speed at which objects fall in air.

At this point in the story it is necessary to give this explanation of the term 'gravity'. Nothing moves unless it is pulled or pushed; pulls and pushes which move things are known as forces. The mechanism by which most forces are exerted can often be seen at work. For example, we can see a crane pulling on a rope when it is lifting a load of bricks; we can see an engine exerting its pull on all its carriages because each carriage is coupled by an iron link to the one in front of it. But there are a few forces which act without a rope or chain or any other solid connection. For example, a magnet attracts a piece of iron; that is, it pulls the iron towards it, but there is nothing solid connecting the two. The iron is pulled by a force called magnetism, a force which is, of course, an invisible one. Another force without any 'solid' connection is the one exerted by the earth which tends to pull objects to the ground and is called the force of gravity (chapter 33).

Knowledge about the force of gravity became most important after the introduction of artillery into warfare (chapter 4), for then problems concerning the flight of cannon balls in air had to be considered. It was obvious to some scholars that there were two forces acting on the moving cannon ball. There was the force produced by the explosion which shot the ball high into the air, and the force of gravity, which, pulling earthwards, brought the ball back to the ground.

The manner in which an object falls in the air had been the subject of study for centuries. Aristotle had written that a heavy body falls to the ground, from a height, much faster than a lighter one, stating that an object whose weight is a hundred times greater than that of another will fall a hundred times faster than it. Galileo questioned this statement and determined to test its truth by actually dropping a heavy and a light ball from a great height at the same time.[2]

He could have found no better place to try this experiment

Galileo and the Leaning Tower

than Pisa, because in that town there is the famous 'leaning tower'. The building was begun in the twelfth century as the bell tower, or *campanile*, of the Cathedral and is about one hundred and eighty feet high with seven tiers or floors, and a belfry. It leans over to what seems a frightening degree, being at the top about fourteen feet out of the perpendicular. For centuries it was supposed that the tower had been purposely built that way; now it is thought that the foundations were laid on wooden piles driven into the boggy ground and that when the tower had been built to a height of about thirty feet it began to settle on one side. Despite the slant it was then decided to finish building it up to its present height. People looking out over the seventh balcony can see straight down to the ground more than a hundred feet below them.

One day in 1590, so the traditional story relates, Galileo climbed the long spiral staircase of the tower up to the seventh gallery, carrying with him two metal balls, one weighing ten pounds and the other weighing only one pound. (Some writers give the weight of one ball as a hundred pounds and that of the other as one pound; others merely state that one ball weighed ten times as much as the other.)

He leaned over the gallery and saw the crowd which had gathered to watch this public experiment. It included members of the University of Pisa – the professors, philosophers and students. All these people knew that Galileo's belief was opposed to that which had been accepted for hundreds of years, and according to one version of the story there were many angry mutterings as the young upstart slowly mounted the tower to try to disprove the beliefs of his elders.

He balanced the two balls carefully on the overhanging edge of the parapet of the gallery and allowed both of them over to begin their fall at the same time. The crowd saw that they kept together in falling through the air and heard a single crash as both hit the ground at exactly the same time. The people were amazed for, in accordance with the long accepted belief, they had expected that the heavier ball would fall much faster than the other and thus hit the ground before it.

Some of the versions of this story give so many details that it is interesting to turn to the first recorded one, which was written in 1654. The main details given there are simply that Galileo demonstrated by repeated experiments from the height of the *campanile* – and in the presence of teachers, philosophers and the whole assembly of students – that heavy bodies falling in air all moved at the same speed; and that these experiments dismayed the philosophers.

This story of Galileo and the Leaning Tower is one of the most familiar ones in the popular histories of science but there are reasons for regarding it as fictitious.[3]

No mention of the demonstration is given in the writings of any one living at the time when it is supposed to have taken place and not even Galileo himself ever once alluded to it in any of his many books. Yet had it really occurred it would have been such a striking event that some one person then alive would surely have mentioned it. The first account of it is found in a biography of Galileo, written by Viviani, a great admirer of his, which appeared in print sixty-four years after the reputed date of the experiment.

There are many instances in the history of science where an admirer has given his hero the credit of doing some outstanding thing which in fact was done by some other person. Viviani might have done so in this case, for it has been well established that others before Galileo had attacked Aristotle's statement that bodies fall at a speed proportional to their weight. It is also well established that a very similar experiment to the one reputedly done by Galileo was made before 1590 by a certain Simon Stevin of Bruges.

Stevin was a brilliant military engineer who became quartermaster general to the army of Holland; he is also known for his mathematical skill, for he was largely responsible for introducing the decimal system into mathematics.

Stevin was assisted in his famous experiment by his friend De Groot. Two balls of lead, one ten times heavier than the other, were dropped together out of an upstairs window on to a plank. The lighter ball did not take ten times longer in fall-

Stevin lets fall two balls

ing than the heavier one, as Aristotle and others had taught; on the contrary, the two balls hit the board underneath the window 'so simultaneously that the two sounds seemed to be one and the same rap'.[4]

This experiment was done in 1587 but there seems to be no

evidence that Galileo knew about it. Viviani may have heard of it and then had the idea of transferring the credit for doing it first to Galileo; and the fact that Galileo lived near such an ideal setting as the Leaning Tower of Pisa could strongly have tempted him to do so.

It appears, therefore, that Galileo was not the first even to think of this experiment. Nevertheless, if he did not actually do the experiment he most certainly taught the result that would have been obtained from it as the following passage from one of his books shows: 'I can assure you that a cannon ball weighing one or two hundred pounds will not reach the ground by as much as a span ahead of a musket ball provided both are dropped from a height of two hundred cubits'.

This passage may well have caused Viviani to believe that Galileo had actually dropped the two balls from the Leaning Tower, which was of approximately such a height.

28. *The Telescope and the Pendulum*

JOHN LIPPERHEIM was a spectacle-maker of Middlesburgh in Holland.[1] One day one of his children when playing in the workshop with two spectacle lenses, happened to place one lens some distance from the other and then took a glance through both at the same time. To his utter amazement he saw that the weathercock on top of the church tower not only appeared to be upside down but also seemed very much larger and nearer to him than ever before. He shouted for his father, who, having looked through the two lenses, decided to make a simple experiment. He fastened one of the lenses to a board and held the other immediately behind it and in line with the weathercock. Then, whilst looking through both lenses, he moved the second lens backwards or forwards until he got the clearest view of the cock.

This is only one of the stories told about the invention of the telescope. A somewhat similar story is told of James Metius, a Dutchman, who, whilst idly amusing himself with his burning glasses (lenses), thought that he would look through a convex lens and a concave one at the same time. So he put the convex one in front of the other and looked through them at a distant object. To his great surprise he saw that the object looked larger and more distinct, and much nearer to him than in fact it could possibly be. But in this case it was the right way up.

In a somewhat similar story, another Dutch spectacle-maker named Jansen is given as the accidental discoverer of the telescope. In the year 1609, Jansen, according to the story, went a step further than the other two Dutchmen. He fixed the two lenses in a tube which could be held easily before the eye. Then he hastened with this new instrument to Maurice, Prince of Orange and Count of Nassau. Maurice was the ruler of the United Provinces (or the Netherlands as we should now say) and was at war with France. He was a brilliant general and immediately realised the great use of such an instrument in military operations. So he ordered Jansen to keep his invention a secret. But it was impossible to hide such a secret for long. Several persons soon began to make and sell the telescope, including, according to one version of the story, Lipperheim. These telescopes magnified about fifteen to sixteen times.

Galileo, the famous Italian scientist, was then living in Venice and learned about this invention. In his own words:

About ten months or so ago a report reached my ears that a Dutchman had constructed a telescope, by the aid of which visible objects, although at a great distance from the eye of the observer, were seen distinctly as if near; and some proof of its wonderful performances was reported, which some gave credence to, but others contradicted. A few days afterwards I received confirmation of the report in a letter written from Paris by a noble Frenchman, Jacques Badovere, which finally determined me to give myself up first to inquire into the principle of the telescope, and then to consider the means by which I might compass the invention of a similar instrument, which after a little while I succeeded in doing, through deep study of the theory of Refraction; and I prepared a tube, at first of lead, in the ends of which I fitted two glass lenses, both plane on one side, but on the other side one spherically convex, and the other concave. Then bringing my eye to the concave lens I saw objects satisfactorily large and near, for they appeared

one third of the distance off and nine times larger than when they are seen
with the natural eye alone. I shortly afterwards constructed another telescope with more nicety, which magnified objects more than sixty times. At
length, by sparing neither labour nor expense, I succeeded in constructing
for myself an instrument so superior that objects seen through it appear
magnified nearly a thousand times, and more than thirty times nearer than
if viewed by the natural powers of sight alone.[2]

In another book he continued the history:

As the news reached Venice that I had made such an instrument, I was summoned before Their Highnesses, The Signoria, and exhibited it to them, to
the astonishment of the whole Senate. Many noblemen and senators,
though of great age, mounted the steps of the highest Church towers at
Venice, in order to see sails and shipping that were so far off that it was
two hours before they were seen steering full sail into the harbour without
my spy-glass, for the effect of my instrument is such that it makes an object
appear as if it were ten times nearer than it actually is.[3]

Afterwards Galileo presented one of his newly made instruments to the Doge and Senate of Venice, together with a docu-
ment in which he explained its structure and the wonderful uses
that might be made of it by land and sea. 'In return for so noble
an entertainment', we are told, 'the Republic, on the 25 August,
1609, more than tripled his salary as professor in the University
of Padua'.

For a month Galileo was besieged from morning to night by
crowds of people eager to have a look through his wonderful
tube. But as well as amusing himself and others by looking at
earthly objects, he turned his telescope towards the heavens and
observed the moon. It must have given him a great thrill when
he saw what no one else had ever seen – hills and valleys on the
moon's surface'. Before long he had 'discovered' numerous
stars, and found that the milky way was made up of countless
tiny stars. But what was probably his most important observation came when he examined the planet Jupiter and noticed
that Jupiter's moons or satellites were revolving round the
planet.

This was a most exciting discovery and made him more certain
than ever before of the truth of the theory of Copernicus. For
Copernicus had taught that the sun did not move across the
heavens, as man had thought for generations, but remained per-

Galileo looks at the moon

fectly still, appearing to move only because the earth itself moved around it. Galileo, having the 'evidence of his eyes' that the moons of Jupiter moved round that planet, felt justified in believing that our moon also moves round the earth. In the same way he felt justified in believing the theory of Copernicus that the earth moves round the sun.

Galileo made a very true forecast when he wrote the following words:

Perchance other discoveries still more excellent will be made from time to time by me or other observers, with the assistance of a similar instrument, so I will first briefly record its shape and preparation, as well as the occasion of its being devised, and then I will give an account of the observations made by me.

In this chapter three men of Holland, Lipperheim, Metius and Jansen, have been mentioned as the discoverers of the telescope. While it is not clear which of the three should be credited with the discovery, there is no doubt at all that the telescope was first made in Holland about the year 1608. Nevertheless it must be emphasised that the Dutchmen used their telescopes to view distant objects on earth, and so to Galileo must go the credit of being the first to use the instrument for the scientific purpose of studying the heavens.

*   *   *   *   *

It is interesting to note, in passing, that Galileo realised that there would be other advantages in using optical instruments; indeed, he came very near to inventing the microscope, for he used his telescope to view small nearby objects. This is his account of what he saw when he directed his telescope on flies. 'I have seen flies which look as big as lambs and have learned that they are covered over with hair and have very pointed nails by means of which they keep themselves up and walk on glass, although hanging feet upwards, by inserting the point of their nails in the pores of the glass'.

His conclusions about the way a fly walks on glass were of course quite wrong and his observation of the fly was by no means perfect. A telescope is not a suitable instrument for magnifying nearby objects, partly because of the small field of view it affords. In any event it was not used long for this purpose for the microscope itself was invented only a few years after the telescope. Soon afterwards, according to Macaulay the historian, it became quite the fashion for English gentlefolk to look at flies and other objects through the microscope.

*   *   *   *   *

Another well-known story about Galileo tells of his discovery of the pendulum.[4] Galileo, we are told, when a student of nineteen years of age, was praying in the Cathedral at Pisa one day in

1583, but soon wearied of doing so. He dreamily fixed his eye on the beautiful lamp designed by Maestro Possenti which hung from an arch. The lamp had just been lit by a verger and it swung to and fro when he released it; at first the swings – or oscillations as scientists call them – were considerable, but gradually they grew smaller and smaller until the lamp came to rest.

Galileo watches the swinging lamp

Galileo noticed that the time for a complete swing seemed to be the same whether the oscillation was a large or small one. He was then studying medicine and knew that a man's pulse beats time regularly in normal cases. So he decided to check his idea by counting the number of times his own pulse beat whilst the lamp was making a certain number of swings. In this way he proved that each completed swing of the lamp, large or small, did take the same time. This gave him the idea of making what has now become known as a simple pendulum.

It consisted of a long piece of string, held hanging downwards, with a small ball fastened to its lower end. The ball, when set swinging, behaved like the lamp: every complete oscillation, whether large or small, took the same time. But he also discovered that by altering the length of the string he could alter the rate of oscillation.

Galileo then thought of a method of using a simple pendulum to measure the pulse rate of man and invented the instrument known as a pulsimeter. This time-measuring apparatus was first used by physicians about the year 1607 and proved of great value to them.

Many years later, in 1641, the idea occurred to him of making a clock whose movements could be regulated by a pendulum. Such a clock, he believed, would keep much better time than the imperfect clocks then in use. Galileo was then a blind old man, so he sought the aid of his son Vincenzo. Vincenzo, we are told, was a very clever mechanic and made to his father's specifications first drawings and then models. But Galileo fell ill and did not recover, so his work on the pendulum clock was not completed.

*   *   *   *   *

An English authority on Galileo makes this comment about this story of the pendulum.

Whether this be only a pretty fable, like that of Newton and the apple, cannot now be decided; but it is, at least, certain that Possenti's lamp was not the one which Galileo observed, since it was not made until 1587 and was only hung in its present place on the 20th December of that year.[5]

It may well be, of course, that Galileo did see a swinging lamp, even if not the one made by Possenti, and there is some evidence that his son did actually construct the pendulum mechanism in 1649 with the aid of a locksmith named Balestri.[6]

The son, Vincenzo, however, died soon afterwards, just a few years before a Dutch scientist named Huygens described a pendulum clock which he had designed in 1658.

29. And Yet it Moves

IT IS NOT surprising that early man believed that the sun moved across the heavens, for it was a movement he could see apparently taking place day after day with great regularity. The belief was firmly established in men's minds even in the days of Solomon, King of Israel, who said: 'The sun also riseth and the sun goeth down and hasteth to the place where he rose';[1] and of Joshua who gave the order, 'Sun stand thou still upon Gideon'.[2] Neither would have spoken thus if he had believed that the sun did not move across the sky. Because of these two passages of Scripture, and of others of a similar kind, the Church taught that the sun moved whilst the earth stood still; and the teachings of the early and mediaeval Church on any subject, religious or secular, had to be accepted by everyone on pain of severe displeasure or dire punishment, often death.

In the year 1543 appeared a book in which it was stated that the sun stood still whilst the earth revolved around it. It is not surprising that the theory came as a great shock to many educated people. Copernicus, the author of the book (page 32), knew that his theory, which contradicted the general belief, would be extremely unpopular, and he feared the anger of the Church. So he delayed publication time after time, with the result that the book appeared in print almost on the day of his death.

Some years later an Italian scholar named Bruno accepted the Copernican theory, and wrote learned books supporting it. For this he incurred the grave displeasure of the Church. He was summoned before the Holy Inquisition and imprisoned. Later, in the year 1600, he was excommunicated and then burned at the stake as a heretic.

The Reformation, as is well known, had split the Christian Church into two parties. But they were united at least in their belief that the sun moved. The Catholics regarded disbelief of this as a mortal sin; Luther, the leader of one body of Protestants, dubbed Copernicus, 'this fool, who wants to turn the whole art of astronomy upside down;' and added, 'but as the Holy Scripture testifies, Joshua ordered the sun to stand still, not the earth', whilst Calvin, the leader of another body of Protestants, asked, 'who will venture to place the authority of Copernicus above that of the Holy Spirit; is it not written in Psalm XCIII, that "the world also is established that it cannot be moved"'.

It is easy for us to criticise the action of those leaders of religion, but we should remember that the acceptance of these new ideas would have meant the admission that most of the teachings of the Church had been based on false ideas for centuries; the whole of the churchman's intellectual world was at stake.

Until the year 1609, astonomers had inspected the heavenly bodies by the naked eye, but in that year Galileo first used the astronomical telescope, and his observations of the planet Jupiter and its moons led him to agree with the view of Copernicus (chapter 28).

After the discoveries resulting from the use of the telescope his views gradually became known. Hence, in the year 1616, the Church found it necessary to state that the belief that the sun stood still whilst the earth moved was a false one. Two days after this statement, Galileo was summoned to appear before Cardinals of the Sacred College. It is said that he was then officially warned not to hold, teach or defend such an idea; and that he promised obedience.

There has been some dispute between writers as to whether he was given such a warning or whether he was simply informed that Copernicus's book had been banned by the Church. (A banned book was not to be read by any faithful Catholic.) Be that as it may, Galileo, who was a devout Catholic, made no further public statements about the theory until the year 1630. Then he published his most famous work, which was called *Dialogues Concerning the Two Principal Systems of the World*.

In this book he strongly supported the Copernican theory. Although he sought and obtained permission to print the book from the relevant Catholic authorities, its publication made him many enemies, especially among the Jesuit and Dominican priests. They stirred up so much feeling that before many months had passed the Holy Office appointed a commission to examine the teachings in the book. It reported against them, and Galileo was ordered to appear for trial.

He was then a sick man of seventy and pleaded that he was unfit to travel to the trial, but the authorities insisted on his presence, although on his arrival in Rome they allowed him to remain in a friend's house instead of being imprisoned in the usual way. Little happened at his 'first examination', except that he pleaded that he had written the book 'with good intention'. At his second examination, however, he was probably threatened with torture in the first degree unless he renounced his writings. (This torture, the *territo realiso*, consisted of showing the victim the implements of torture with a detailed explanation of how they worked and what they did.) He confessed, on oath, that his views were false.

Accordingly, on 22 June 1633, a solemn session of the Inquisition was held in the monastery of Santa Maria sopra Minerva in Rome. It was attended by numerous Cardinals (his judges) and high officers of the Church, probably all ceremonially robed. His sins of 1615 were recalled; he was reminded of the promise of obedience which he had made in 1616; and finally this sentence was pronounced:[3]

Whereas you, Galileo, were denounced in 1615 to the Holy Office, for holding as true, a false doctrine taught by many, and for answering the objections produced from the Holy Scriptures by glossing the said Scriptures according to your own meaning, therefore that Holy Tribunal decreed as follows:
First: The proposition that the sun is in the centre of the world and does not move is absurd, philosophically false and formally heretical because it is expressly contrary to the Holy Scripture.
Second: The proposition that the Earth is not the centre of the world, nor immovable, but that it moves is also absurd, philosophically false and theologically considered at least erroneous in faith.
But being pleased at that time to deal mildly with you, it was decreed in the Holy Congregation that His Eminence the Lord Cardinal Bellarmine should enjoin you to give up altogether the said false doctrine. Accordingly,

you were commanded in future neither to defend or teach it in any manner, neither verbally nor in writing and upon you promising obedience you were dismissed.

After giving this sketch of what had happened on the former occasion in 1616, the pronouncement then continued by stating that Galileo had confessed to writing a certain book wherein he had defended his former opinion. This, the sentence continued,

is a really grievous error because no opinion can by any means be provable after it has been declared and determined to be contrary to Divine Scripture. Therefore, having seen and maturely considered the merits of your cause, with your confessions and excuses and everything else which ought to have been seen and considered We have come to the final sentence against you:

We pronounce, say, judge and declare that you, the aforesaid Galileo, have rendered yourself suspect of heresy by this Holy Office; that is to say that you believe and hold a doctrine which is false and contrary to the Holy and Divine Scriptures after it has been declared and determined contrary to Holy Scripture; consequently you have incurred the censure and penalties enjoined and promulgated in the sacred canons and elsewhere against such offenders; from which it is our pleasure that you should be absolved, provided that you do first, with a sincere heart, and faith unfeigned, abjure, curse and detest before us, the said errors and heresies, and every other error and heresy contrary to the Catholic and Apostolic Church of Rome, in the form that shall be given to you by us.

We further decree, that the book of Galileo Galilei shall be prohibited by a public edict; and We condemn you to be formally imprisoned in this Holy Office for a period determinable at our pleasure; and by way of salutary penance We order you, during the three years ensuing to recite, once a week, the seven penitential psalms[4] reserving to ourselves the power of moderating, changing, or wholly or in part removing, the aforesaid punishment and penance.

Galileo was then compelled to kneel and abjure as follows:

I, Galileo, son of the late Vincenzio Galilei, a Florentine, aged seventy, being brought to judgement, and on my knees before you, the Most Eminent and Most Reverend the Lord Cardinals, Inquisitors General of the Universal Christian Commonwealth, against heretical depravity, having before my eyes the most Holy Gospels, which I touch with my hand; do swear that I always have believed, and do now believe, and with the help of God will in future believe, all that the Holy Catholic and Apostolic Roman Church doth hold, preach and teach. But, despite having been judicially ordered and commanded, by this Holy Office to forsake that false opinion which holds that the sun is the centre of the world and immovable, and having been forbidden to hold, defend or teach the aforesaid

29. *And Yet it Moves*

false doctrine, I have nevertheless written and printed a book in which I dealt with the said doctrine. .... I have, therefore, been judged grievously by the said Holy Office as vehemently suspected of heresy, that is to say, that I have held and believed that the earth is not the centre but moves round the sun.

Being, therefore, willing to remove from the minds of your Eminences, and of every Catholic Christian, this vehement suspicion rightly entertained towards me, I do, with a sincere heart, and faith unfeigned, abjure, curse and detest the aforesaid error and heresy, and, in general, every other error and sect contrary to the Holy Church; and I swear, that for the future, I will never more say or assert, either by word or writing, anything to give occasion to such suspicion, but that, if I shall know any heretic or any person suspected of heresy, I will inform against him to this Holy Office or to the Inquisitor and Bishop of the place in which I shall be.

Moreover, I swear and promise, that I will fulfil and wholly observe all penances which have been, or shall be, laid on me by this Holy Office. But if, which God forbid, it shall happen that I shall act contrary, by any words of mine, to my promises, protestations and oaths, I do subject myself to all penalties and punishments which have been ordained and published against such offenders by the sacred canons, and other constitutions, general and particular. So help me God, and his Holy Gospels, which I touch with my hands.

I, the above Galileo Galilei, have abjured, sworn, promised and bound myself as above: and in testimony of these things have subscribed, with my own hand, this writing of my abjuration which I have repeated word for word at Rome, in the Convent of Minerva, this 22nd day of June, 1633.

*   *   *   *   *

It is often stated that Galileo was compelled to wear a hair shirt during the ceremony, but there seems to be no authentic statement about what he actually wore: a contemporary artist showed him in ordinary apparel.

Tradition has it that as soon as Galileo rose to his feet he was filled with remorse at having denied that the earth moved for 'his conscience told him that he had sworn falsely'. He looked at the ground, and, stamping his foot, said 'E pur si muove' ('and yet it moves', or, 'it moves nevertheless'.)

The phrase has passed into the history of science and is frequently quoted. It would seem most unlikely, however, that he said such words in the presence of his judges, for he was then a sick, worn out old man who had just gone through an experience

Galileo before the Inquisition

which would have crushed many a fit and young man. Moreover, his judges would have punished him most severely for such an utterance, which would have been 'contempt of court'.

The earliest known mention of the phrase in a printed book seems to be the following one which was made in 1757 below a portrait of him:

This is the celebrated Galileo who was in the Inquisition for six years and put to the torture, for saying that the earth moved. The moment he was set at liberty, he looked up to the sky and down to the ground, and, stamping with his foot, in a contemplative mood, said *E pur si muove*, that is, 'still it moves', meaning the earth.[5]

It is likely that Galileo, if he spoke these words at all, did so out of court and not in it. Indeed, it is quite possible that he did actually use the phrase when he left court, for he was then in the company of a few of his old friends. Some confirmation comes from an old portrait of him which, in 1911, was taken out of its frame. On the margin previously covered by the frame are a few drawings which could have been deliberately hidden from view. They depict the earth moving round the sun with the words: *E pur si muove*. The picture was painted, probably in 1646, by a Spanish artist at the command of the man with whom Galileo stayed after his sentence.[6]

Few people now believe that Galileo was tortured before he appeared before his inquisitors, although he was probably threatened with torture of the first degree as a matter of routine. The sentence he served was a slight one. He remained in the custody of the Inquisition for two days and was then put under 'house arrest' in the home of a friendly Archbishop. He remained there for several months until he was allowed to return to a house in Florence, where he spent the rest of his life in 'close retirement'.

30. *The Romance of the Barometer*

UNTIL THE middle of the seventeenth century, scientists believed that Nature had a horror of a vacuum. This belief that 'Nature abhors a vacuum' was the basis of their explanation of the working of the pump.

A pump consists essentially of a long pipe, one end of which dips into the water which has to be pumped, whilst the other end is connected to a barrel or cylinder. When the handle of the pump is moved up and down a partial vacuum is formed in the cylinder. Then, said the early scientists, because Nature had such a dislike of a vacuum she got rid of it immediately by 'sending' water up the pipe to fill the empty space.

A traditional story relates that in the year 1640 the Grand Duke of Tuscany decided to sink a well in the grounds of his ducal palace. The workmen had to dig much further down than was customary and did not reach water until they had got to a depth of about forty feet. A pump was erected, with its pipe dipping into the water; and then the men tried it. But to their amazement no water came out of the spout no matter how hard they worked the handle up and down. The men thought that there must be something wrong with the pump; but a careful inspection showed no fault.

This strange occurrence was reported to the Duke, and he, like his men, could not understand why the pump failed. In those days many rich men such as the Duke became the 'patrons' of well-known scientists; that is, they paid a salary to the scientist to enable him to continue his scientific work without having to do much other work to earn a living. Many years before the failure of the pump Galileo had been appointed the Grand Duke's 'philosopher and mathematician extraordinary'. The Duke therefore consulted him about the problem.

It was seen that the water rose to a height of 'eighteen palms' (about thirty-three feet) up the pipe of the pump and no further. Galileo 'explained' this by stating that whilst Nature did not like a vacuum its horror ended when water had risen to a height of eighteen palms inside one. But he himself must not have been thoroughly satisfied with this explanation for, being then an old man, he asked one of his young and promising pupils called Torricelli to investigate the problem.[1]

Torricelli, believing that a pump could not lift a heavy liquid as high as it did a light one, decided to use mercury in his investigations, for mercury is thirteen and a half times as heavy as the same volume of water. He therefore expected that the maximum height which mercury could be lifted by a pump would be thirty-three feet divided by thirteen and a half, that is, thirty inches. The big advantage of using mercury instead of water would be that he could then use a tube about a yard long, which would be a convenient size, instead of one at least thirty-three feet long.

He obtained a glass tube of this length which was sealed at one end. First he filled it completely with mercury; then he closed the open end by putting his thumb over it; and finally he turned it upside down and submerged it in a bowl of mercury so that the open end was below the surface. When he removed his thumb from the open end a column of mercury about thirty inches high was left standing in it; and at the top of the tube, where some of the mercury had been, was an empty space (which later became known as the Torricellian vacuum).

Long before the time of this experiment Galileo had shown that air has weight, as have all substances. Hence Torricelli

Torricelli with his tube

concluded that the weight of air acting on the surface of the mercury in the bowl resisted the escape of the mercury from the tube. When the weight of the mercury which was left in the tube was 'balanced' by the thrust of the air on the surface of the mercury in the bowl no more mercury could escape into the bowl.

Torricelli then gave a correct explanation of the failure of the pump. The air, he said, pressed on the surface of the water in the well with sufficient force to send it 13½ times 30 inches, say 33 feet up the pipe of the pump, but it could not force it to go any higher.

The experiment had done more than solve the problem of the pump's failure; it had shown a method of measuring the pressure of the air. Soon Torricelli's tube inverted over mercury became known as a barometer; and even today we still state the pressure of air by saying that it will hold up a column of mercury which is so many inches in height.

About the year 1644 the fact that air exerts a pressure came to the knowledge of a young French scientist named Blaise Pascal when he was living in Rouen. He pondered over the statement which had been made that 'we live at the bottom of a sea of air which undoubtedly has weight'. If this were true, he reasoned, then the less the height of the air above us, the less should be the weight of air pressing on us. Hence if a barometric tube (that is, Torricelli's apparatus) was carried to a great height, say up a tall tower, the length of the column of mercury in the tube should decrease.

He decided to see if this was indeed the case by carrying such a tube up a church tower. But although he noticed a slight difference in the height of the mercury, he realised that the tower was not tall enough to yield any conclusive results. He then thought of his native mountains. He came from a village called Clermont, about two hundred miles south of Paris. This village lies at the foot of the Puy de Dôme, a mountain rising to a height of three thousand feet.

Pascal was then a sick man and had been ordered by his doctor not to take any vigorous exercise, so he persuaded his brother-in-law, M. Perrier, who lived at Clermont, to do the experiment for him.[2]

On 19 September 1648 the summit of the Puy de Dôme could be seen appearing through the clouds at about five o'clock in the morning. So Monsieur Perrier decided to make the experiment that day. He called together his friends and by eight in the morning five men, all distinguished in their respective professions, and all interested in science, were ready to make the ascent.

M. Perrier on the Puy

M. Perrier had provided himself with two glass tubes about four feet long and sealed at one end, two bowls, and about sixteen pounds of mercury. At the foot of the mountain he did Torricelli's experiment with one of the glass tubes and some mercury; he found that the column of mercury in the tube measured 26·4 inches in height.

He repeated this experiment with the other tube and satisfied himself and his companions that the column of mercury was the same height in each tube.

The five men then set out for the summit of the Puy de Dôme, leaving behind one of the inverted tubes in the care of a friend who had offered to take readings of the height of the column at intervals throughout the day.

The summit, about three thousand feet above the starting point, was at last reached, and, on doing Torricelli's experiment, the men found that the mercury stood at a height of 23·2 inches. The column was therefore 3·2 inches shorter than it was at the starting point. It is said that, although they had expected some difference, the reading differed so much from the one taken at the foot of the mountain that they could hardly believe their own eyes and decided to repeat the experiment in various ways and in different places on the mountain top. Thus they repeated it inside a small chapel built on the summit, and again and again in the most exposed places they could find. They even waited until a fog descended on the mountain and again repeated it. But each time they found that at the summit the height of the column was 23·2 inches.

They then began the descent and when they had reached a spot almost half way down the mountain side they decided to repeat the experiment. They found that the height of the column was 25 inches. Upon reaching the starting point they checked the reading on the tube which had been left there and found that it still stood at 26·4 inches.

Next morning the priest of the oratory at the base of the mountain suggested that the experiment should be repeated at the bottom of the high tower of Notre Dame in Clermont and again at the top of the tower. This was done and the difference in the readings was 0·2 inches. The tower was about 120 feet high.

The results of the experiments were communicated to Pascal, who at once repeated them, using a high tower in Paris. He obtained approximately the same results as his brother-in-law.

These experiments clearly demonstrated to Pascal that Galileo's theory that air has weight was correct and that we do live at the bottom of an ocean of air which is pressing on us. They also showed that Torricelli's tube could be used to measure the height or altitude of mountains as well as to record atmospheric pressure.

31. Sixteen Horses versus the Air

OTTO VON GUERICKE was born in Magdeburg in the year 1602, the son of a well-to-do family. After studying mathematics, especially geometry and mechanics, he travelled abroad; for in those days a foreign tour was considered an important part of a gentleman's education. He visited England during the reign of James I and later spent some time at one or two universities on the continent before returning to his home in the capital town of Saxony, a province of Prussia.

In 1618 a bitter war broke out which lasted for thirty years; much of the fighting took place in Germany. Von Guericke played his part in the war and his mathematical training helped him to become a military engineer of some importance. But he was on the losing side, and in 1631 Magdeburg was captured and cruelly sacked, some thirty thousand of its inhabitants being killed and almost all the important buildings destroyed. Guericke, who was a senator of the city, escaped death and later helped to rebuild the city. He was subsequently made the burgomaster (or mayor), an office which he held for thirty-five years.

Although his civic commitments had made his life a very busy one, von Guericke found time for his hobby of scientific research. He knew that Galileo had shown that air had weight and was greatly interested in the work done by Torricelli, and having a sense of humour as well as being most ingenious, made for himself a novel water barometer which gave him much pleasure and amusement. This barometer reached from the ground to the roof of his house. It consisted of four brass tubes joined to make one tube about ten yards long. An elongated upside down flask was sealed to the top of this long tube whose lower end was submerged in a large tube containing water. The barometric height of the water in the tube was about thirty-two feet and there was the usual Torricellian vacuum in the elongated flask.

Von Guericke put a wooden figure shaped like a man – a

maniken, as it was called – in this water barometer, so that it floated on top of the water in the elongated flask. He then concealed all the lower parts of the tube in such a way that no one could see anything except the glass vessel containing the wooden maniken. Furthermore, he concealed the tube to the height reached by the water only in fine weather. Thus the top of the water, and hence the maniken, showed only in fine weather, being hidden behind boarding when the pressure of the air and consequently the height of the water in the tube was low. The weather maniken, who made his appearance only in fine weather, 'excited amongst the populace vast admiration', we are told, 'and von Guericke, the worthy magistrate, was in consequence suspected by some of the townspeople of being very familiar with the powers of darkness'.[1]

Another of von Guericke's achievements was the designing of an efficient air pump which would give him a vacuum. One of his experiments seems a very simple one. He filled a barrel completely with water and fastened the pump of a fire engine to a tube fixed to the barrel's lower end. This pump, he hoped, would withdraw all the water, leaving an empty space in the barrel. A few trials showed the necessity of sealing every joint between the staves. This he did, but then as the water was gradually withdrawn the task of working the pump became harder. At length three men had to exert their full power to pull the piston out. Finally the wooden staves gave way and the air rushed into the barrel with a loud noise.

Von Guericke then knew that a wooden cask was not strong enough to contain a vacuum so he used instead a hollow ball of copper. But the physical effort needed to work the pump became too great. In a short time even four very strong men could scarcely move the handle.

He then invented his well-known pump which withdrew air and not water from a closed space. Another hollow copper ball was made. Two large cups of copper were used, each of which formed exactly half of the completed ball or sphere, and so was called a hemisphere.[2] They fitted close together, rim to rim, to make a complete hollow globe. Von Guericke wanted this globe to be airtight. So he dipped a leather ring which had the same diameter as the hemispheres into a solution of wax in turpentine. When the ring was withdrawn from this solution and left to dry, all the turpentine evaporated and left the wax in the pores of the leather. The ring was thus airtight and he put it as a kind of washer between the rims of the two cups. One of these cups had been fitted with a stopcock, and a stout ring had been fastened to the outside of each of them. When the cups and washer were fitted together von Guericke had a hollow airtight copper globe about thirty inches in diameter.

He set his men to work with his newly invented air pump until all the air had been pumped out of the copper ball and was then ready to try his experiment. This he did in the presence of only a few friends so that he might be sure that it would work before he gave a public demonstration. The friends gathered together at Regensburg in front of the Reichstag or Houses of Parliament. The experiment was a great success.

In 1651 Emperor Ferdinand III heard about it and commanded von Guericke to demonstrate it in front of him; so

Sixteen horses at tug of war

shortly afterwards the event pictured here took place. The Emperor and some of his courtiers can be seen watching a tug-of-war such as they had never seen before.

Eight strong horses were harnessed to one cup and eight to the other, for, as von Guericke wrote, the experiment was designed 'to show that, owing to atmospheric pressure, two hemispheres can be so firmly united that they are not torn apart by sixteen horses'.

The horses tugged and tugged without success. Finally, by exerting every ounce of power they possessed, they managed to pull the hemispheres apart. When the cups parted the royal spectators were greatly alarmed, because, to use the words of von Guericke, 'when, in the end . . . the horses parted the cups, this was accomplished by a bang like that of a gun-shot'. (The noise was caused, of course, by the air rushing into the empty cups.)

After the Emperor and courtiers had thus seen how difficult it was to pull the two cups apart von Guericke showed them a very easy way of doing it. He loosed the horses, put the two cups together again to make a hollow globe from which his assistants pumped out all the air. Then he simply turned the tap. The air rushed into the globe and von Guericke pulled the two cups apart without the slightest effort. This was because the air inside them was then pushing on the inner surface of the globe with as much force as the external air was pushing on the outer surface and so the two pressures cancelled out.

He then made a calculation to find the pressure of the air on the outer surface of a larger globe, a yard in diameter, and found that it would be so great that twenty-four horses would not be able to pull the two halves apart. So he made another experiment using larger cups than before and employing twenty-four horses instead of sixteen. The horses failed to part the cups; but von Guericke managed it simply by turning the tap.

32. Newton and the Apple

IN THE year 1664, Isaac Newton, then in his early twenties, was a member of Trinity College, Cambridge, where he was studying mathematics. In that year many hundreds of people living in London were dying of the plague, a dreadful disease which later spread to other parts of the country, especially during the hot summer months of 1665.

The disease was very infectious, so many people sought safety in the small villages of the countryside where the risk of infection was thought to be less than in the crowded towns. Newton could have found few places where the risk seemed smaller than at his mother's home in the tiny village of Woolsthorpe, six miles from Grantham in Lincolnshire. So he left Cambridge and spent the next two years or so with his mother.

She lived in a house with a pleasant garden in which Isaac spent many hours of study. Later in life, he wrote that he had thought more about mathematics and science during the two plague years than at any other time of his life, being then, as he said, 'in the prime of his age for invention'. It was during this time that he discovered what is now an important branch of mathematics (the differential calculus) as well as many new facts about light and some of the laws governing the force of gravity.

The most familar story about him concerns the last-named of these topics. It is as follows:

One day, as Newton was sitting in his mother's garden at Woolsthorpe, he saw an apple fall from a tree. This set him thinking of the reason why it should fall straight down to the ground. Why, for example, did it not go upwards or sideways, instead of dropping perpendicularly to the ground. He came to the conclusion that the apple fell downwards, when the stem broke, because some force was pulling it to the ground.[1]

Thus, so the story concludes, this chance-observation led him 'to discover the force of gravity'.

Newton and the falling apple

The first mention of this incident is probably the brief statement of Robert Greene in his book on forces, which appeared in 1727. In this book he dealt with Newton's idea about gravitation and commented, 'This well known idea takes its origin, so it is said, from an apple. This I received from that very clever, learned and also excellent man, who was moreover my very dear friend, Martin Folkes, Knight, a very worthy Fellow of the Royal Society, whom I mention here to do him honour'.[2]

A few years later the Frenchman Voltaire also mentioned it in his *Letters Concerning the English Nation* (1733), in the following words:

Newton being retired in 1666, upon account of the plague, to a solitude near Cambridge, as he was walking one day in his garden saw some fruit fall from a tree, and he fell into a profound meditation on that gravity, the cause of which has been so long sought, but in vain, by all philosophers, whilst the vulgar think there is nothing mysterious about it.[3]

Some years later, Voltaire acknowledged that Newton's half-niece, Mrs Conduit, had told him of the incident; and it may well be that she had also told Martin Folkes.

32. *Newton and the Apple*

In the next century many philosophers refused to believe that such a simple incident as the fall of an apple could have anything to do with Newton's brilliant work on gravitation. It was noticed that many writers of his day did not mention the incident, as they most probably would have done if they had heard of it. For example, Fountenelle, who wrote the eulogy on Newton's death, did not mention it, despite the fact that he got much of his information from Mrs Conduit. Pemberton, another contemporary, merely wrote, 'The first thoughts which gave rise to his Principia he (Newton) had when he retired from Cambridge in 1666 on account of the plague. As he sat in a garden he fell into a speculation on the power of gravity'.[4] Another of Newton's contemporaries, Whiston, did not mention it in his book on Newton; neither did Newton's main biographer, Sir David Brewster.

Some writers went further than registering disbelief of the story; they ridiculed it. Thus Hegel, a German, described it as 'that very lamentable story of the apple which fell before Newton's eyes', and then added, 'those who are delighted with the story must have forgotten all the evils an apple has brought to the whole world, including the fall of man and the fall of Troy. The apple is a bad omen for the philosophical sciences'. Gauss, a fellow countryman of Hegel's, gave an amusing version of the traditional story.

The history of the apple is absurd. Whether the apple fell or not, how can anyone believe that such a discovery could in that way be accelerated or retarded? Undoubtedly the occurrence was something of this sort. There comes to Newton a stupid, importunate man who asked him how he hit upon his great discovery. When Newton had convinced himself what a noodle he had to do with, and wanted to get rid of the man, he told him that an apple fell on his nose. This made the matter quite clear to the man and he went away satisfied.[5]

*   *   *   *   *

That part of the story which states that the sight of a falling apple led him to discover the force of gravity can be readily dismissed, for many men before him knew something about this force. For example, Galileo, who died in the year Newton was born, contributed much to man's knowledge of gravity (chapter 27). But the sight of its fall could have inspired Newton to study gravitation more thoroughly than anyone before had done.

The possibility that he did this became strengthened when a story of Newton's life, written by Dr Stukeley, his physician, came to light. (It had lain in manuscript form practically unnoticed for nearly two hundred years.) In it the doctor stated that he wrote from matters of his own knowledge and 'not from hearsay', and gave the following account:

> On 15th April, 1726, I paid a visit to Sir Isaac, dined with him and spent the whole day with him alone. After dinner, the weather being warm, we went into the garden and drank thea (tea) under the shade of some apple trees, only he and myself. Amidst other discoveries, he told me, he was just in the same situation, as when formerly the notion of gravitation came into his mind. It was occasioned by the fall of an apple as he sat in a contemplative mood. Why should the apple always descend perpendicularly to the ground? thought he to himself. Why should it not go sideways or upwards, but constantly to the earth's centre? Assuredly, the reason is that the earth draws it.[6]

Evidence such as this cannot well be disputed and it does seem that the sight of an apple falling to the ground did set Newton thinking about gravity.

In the late eighteenth century, one particular tree in the garden at Woolsthorpe was earmarked as the one from which the apple fell. By the year 1820 it had become so decayed that it had to be cut down, but its wood was carefully preserved and some of it was used to make a chair which is still in existence.[7]

In 1951 the *Lincolnshire Echo* reported that descendants of this famous tree were still growing. It would appear that grafts had been taken from the original tree and had been sent to a leading fruit research station, where they had been re-grafted. In this way new trees were produced, one of which was sent to America. Some of these trees yielded apples which have been identified as 'The Pride of Kent', a favourite cooking apple of Newton's day.

*   *   *   *   *

Sir Isaac has been the subject of many anecdotes, one of which tells of the loss of some of his important papers in a fire.[8]

32. Newton and the Apple

Newton, the story runs, when at Trinity College, Cambridge in 1694 (he was then fifty-one years old), was engaged in writing a book describing the experiments he had done during the previous twenty years. One winter's day, before attending early morning service in the college chapel, he accidentally shut his dog, Diamond, in his room. On his return he found that the dog had knocked over a lighted candle and set fire to the papers on which he had written an account of his experiments. At first the sight of his papers – representing the work of years – in ashes seemed to affect him but little and he merely commented, 'Oh Diamond, Oh Diamond, thou little knowest the mischief thou has done'; he did not even punish the dog.[9] But before long, the grief caused by this great loss affected his health so seriously that for a time he well-nigh lost his reason.

*   *   *   *   *

It seems well established that a fire did break out in Newton's rooms at Trinity College about this time and that Newton lost some of his valuable papers in it. About the same time also, he did have a serious illness with marked insomnia, but whether this was caused by the loss of his papers is not known.

The part played by the dog Diamond is very doubtful. Newton's clerk stated, in another connection, that his master disliked dogs and cats and kept neither in his chambers. If this statement is accepted it must be concluded that if any dog did knock over the candle it was not Newton's animal.

A lighted candle, however, was the cause of the fire, according to information from the one person most likely to know what really happened, namely Newton's physician, Dr Stukeley; and he wrote, 'Dr Newton tells me that several sheets of his *Optics* were burnt, by a candle left in his room. But I suppose he was able, by a little pain, to recover again; or, if there be any imperfection in that work, we may reasonably suspect it was owing to this accident'.[10] Stukeley, it will be seen, made no mention of the dog.

33. Some Early Electrical Experiments

THE ANCIENT Greek philosophers knew that when the sub-stance called amber is rubbed it attracts small pieces of straw and dried leaves, but little use was made of this knowledge until Dr Gilbert did his famous experiments during the reign of Elizabeth I.

He discovered that other substances behaved like amber and called all of them 'electrics' after the Greek word for amber, which is 'elektron'. Gilbert did many experiments with amber and also with magnets and some were so interesting that he was commanded to demonstrate them to the Queen. His work laid a good foundation, and rapid progress was made during the eighteenth century in this new branch of science, which later became known as 'magnetism and electricity'.

Some of the most interesting of the eighteenth-century experi-ments were done by Stephen Gray of Charterhouse. Between the years 1720 and 1730, working at home with very simple ap-paratus, he proved that some substances conduct electricity whereas others do not.

In these experiments Gray obtained an electric charge by rubbing a glass tube, about a yard long and an inch in diameter. The electrified tube attracted small feathers and pieces of metallic foil; it also imparted a prickling sensation to anyone touching it.

The main piece of apparatus for one of the first of his many noteworthy experiments was a long line of stout sewing thread, called pack thread. It was passed through loops of silk hanging from the ceiling and so was suspended in the air. Gray then held his electrified glass tube to one end of the thread and put small feathers close to the other end. The feathers were attracted to that end of the thread. He therefore knew that the electricity had passed from the glass and along the whole length of the line of thread – a distance of three hundred feet.

33. *Some Early Electrical Experiments*

Gray later substituted loops of brass wire for those of silk, but on repeating the rest of the experiment, discovered that the electric charge had not travelled to the end of the line of pack thread. Evidently loops of brass differed from loops of silk cord in so far as when electricity 'came to the wires that supported the line it passed by them to the timbers to which each end of them is fixed'. That is, the electricity had been conducted along the brass wires to the ceiling and there became 'lost', whereas his first experiment showed him that electricity is not conducted by silk cord.

The result of this experiment led him to make a series of experiments on electrical conduction and insulation. In these he used common household articles. For example, he tied one end of a silk cord to the ceiling and the other end to the handle of the kitchen poker. Next he placed an electrified glass tube to the handle of the poker and placed feathers near the other end. The feathers were attracted. Hence he knew that an iron poker conducts electricity.

Other articles suspended in a similar way were the fire tongs, a copper kettle, a sirloin of beef, a metallic tea pot, a red-hot poker, a map of the world and a cockerel. Each was electrified at one spot and a test was made to see whether the electricity had travelled along the article. These experiments enabled Gray to classify numerous substances as conductors or insulators of electricity.

As will now be obvious to the reader, Gray was a most ingenious man. When he wished to find out whether or not the human body conducted electricity, he decided to make use of his foot-boy. This foot-boy, 'a good stout lad', was suspended from the ceiling by two very strong lengths of silken cord with loops at the lower ends. The lad was placed face downwards, in a horizontal position, with his feet in one loop and his shoulders in the other. He was then lying suspended in air. Gray rubbed a glass rod (thus charging it) and put it to the soles of the boy's feet. He touched the boy's head with his finger and felt a prickling sensation; thus he knew that the electricity had passed along the length of the boy's body.

In another series of simple experiments, Gray held a metal rod in one hand and then brought a charged glass rod as near to this rod as he could without actually touching it. Electricity flowed across the small gap between the two rods in the form of sparks,

Abbé Nollet and the boy

and he heard a snapping or crackling noise like that of a slight explosion.[1]

Today most of us are well acquainted with the facts underlying these effects, but in those days they were novel; and for many years after Gray's time electricity was largely a matter of sparks and shocks and had no useful purpose.

Gray's experiments on the 'good stout lad' attracted the attention of a French scientist, the Abbé Nollett, who decided to repeat them. He also suspended a boy on silken cords. The boy can be seen in the illustration holding one hand over a table on which small pieces of metallic foil had been placed. When the boy was touched by a charged rod the onlookers were thrilled to see the foil jumping up from the table to his hand.[2]

In another experiment the Abbé suspended a fellow-scientist horizontally and brought a charged glass rod to his feet. The Abbé then placed his hands about an inch above his friend's face. Immediately there was a crackling noise and both men felt a little pain like the sudden prick of a pin. When they repeated the experiment in a dark room they saw 'sparks of fire' jumping from the friend's face to Nollet's hand. Although both scientists had expected something like this, the novelty was so great that the Abbé later said that he would never forget the excitement which the first electric spark ever drawn from the human body produced in him.

Until about the year 1740, most experimenters obtained electricity by charging a glass rod or a tube by hand, for, although electrifying machines had been invented some years before, they only gradually came into use. A typical machine consisted of a glass cylinder, which rested on pivots and had a handle attached to it, and a silk cushion which was fixed so that it gently pressed against the glass. When the handle was turned the cylinder rotated and, in rubbing against the cushion, became charged with electricity by friction. The user had then to connect the machine to the article to be electrified by placing a long tube of metal so that it touched the machine and the article. Some scientists used a gun barrel for this purpose.

In 1746 Pieter van Musschenbroek, a professor from Leyden,

observed that an electrified article soon lost its charge on standing and thought that the loss could be prevented by completely
enclosing the electrified article by a non-conductor. So he decided to test his idea by electrifying a small quantity of water in a
glass jar.

He fastened a length of brass chain to one end of a gun barrel,
the other end of which he put in contact with an electric machine.
Cunaeus, another scientist who was assisting him, then held a jar
of water in such a way that the brass chain dipped into the water,
whilst Musschenbroek turned the handle of the machine.

Electricity flowed along the barrel, down the brass chain and
into the water. After a time Cunaeus, who was still holding the
jar on the palm of his hand, happened to touch the barrel with
his other hand. All at once he was struck as if by lightning. His
arms and legs were paralysed for a time. But after many hours he
recovered. Shortly afterwards Musschenbroek sent an account of
the incident to a well-known French scientist, remarking 'For the
whole kingdom of France I would not take a second shock'. He
advised his correspondent that the experiment was so terrible
that on no account should it be attempted.

Despite this most unpleasant experience Musschenbroek and
his colleague had made the very important discovery that electricity could be stored in a jar of water.

In a short time the hanging brass chain had been replaced.
The bottle was corked and through the cork was placed a brass
rod which had a knob at the top and short brass chain hanging
from its lower end into the water in the jar. It was charged by
putting the brass knob in electrical contact with the machine.
(By the year 1748 the water had been replaced by a coating of
metallic foil put around the inner surface of the bottle; and the
outer surface had been similarly coated to the same height, as the
illustration shows. (In the left-hand jar the part of the rod which
is hidden by the foil is shown by dotted lines.)

There are, however, other versions of the discovery of the jar
and it would seem that von Kleist, another scientist, also discovered it independently of Musschenbroek about the same time
The jar, however, became known as the Leyden jar.

33. Some Early Electrical Experiments

Leyden Jar

The Leyden jar had to be handled with great care when charged with electricity. Any one who held it on the palm of his hand and touched the knob got an electric shock which could be a very severe one when the jar was fully charged. Sparks and crackling noises could be produced by putting one end of a wire to the bottom outer surface and the other end so that it almost touched the brass knob.

The announcement of Musschenbroek's experiment created great excitement and amazement. Crowds of people flocked to inspect this 'prodigy of nature and philosophy' and wished to know all about the wonderful bottle. Some unscrupulous men disguised themselves as magicians and travelled from fair to fair mystifying the peasantry with their crude experiments and their production of sparks and shocks from a bottle.

Scientists, however, appreciated the great contribution made by the jar because its discovery came at a time when the new subject of electricity was being keenly studied. In particular the Abbé Nollet, the French scientist, did many experiments with it. These were designed to determine the distance through which electricity could be transmitted, the kind of substances through which it would travel, and the rate at which it moved. Two of his experiments were performed before distinguished audiences.

In the presence of the King of France and his courtiers, a

company of one hundred and eighty soldiers of the Royal Guard joined hands and formed a ring with a small gap in it. The soldier at one side of the gap grasped a charged Leyden jar by the outer coating and, when all the soldiers were standing motionless, the soldier just across the gap was ordered to touch the knob of the jar quickly. He did so. At that instant the whole company received such a shock that all of them sprang in the air as one man. Never had soldiers obeyed a command so quickly and so simultaneously![4]

Shortly afterwards Nollet did another public experiment, this time at the Grand Convent of the Carthusians in Paris. There, all the monks formed a ring said to be over a mile long. An iron wire was held between every two of them, and there was a short gap between each end of the line, as before. The monk at one end of this gap grasped a charged Leyden jar by its outer coating. At a signal the monk at the other end touched the knob. Immediately the whole company felt the shock and sprang into the air.

In England a few prominent people formed a committee to witness and report on some new public experiments of this kind. On the 14 July 1747 they met on Westminster Bridge, near the Houses of Parliament. A wire had been laid all across the bridge, which is about a quarter of a mile long and continued down to the water's edge on both sides of the river. On one bank stood a man holding a charged Leyden jar by its outer cover with one hand and dipping an iron rod in the river with the other hand. The jar was attached to the wire. Another man stood opposite to him on the other side of the water, close to its edge. He, too, was holding an iron rod in one hand, and in the other he held the end of the wire.

At a given signal the second man dipped the iron rod into the water. Immediately the two men jumped simultaneously: they had received an electric shock. The electricity, in an instant, had passed from the jar across the length of wire over the bridge, through the man holding the iron rod, down this rod into the water, across the quarter-mile width of the river water, through the iron rod to the man on the opposite bank, and back to the jar.[5]

The effect of the discovery that electricity could pass like a flash across a wide river like the Thames was enormous; indeed it is almost impossible for any of us today to imagine the amazement, bordering on incredulity, with which the information was received.

This experiment was followed by similar ones performed in public which showed that electricity flowed in an instant through a circuit even some miles in length.

These and other experiments attracted great attention, not only in England and other European countries, but also in America, as the next story relates.

34. The Famous Kite of a Famous Statesman

NEWS OF the electrical experiments mentioned in the former chapter spread to the colonists in North America. In the year 1747 a letter from London to the Literary Society of Philadelphia explained some of the recent work on electricity; the writer also sent, as a present, one of the glass rods commonly used in the electrical experiments then being done in London.

Benjamin Franklin, a forty-year-old printer of the town, became deeply interested in this new subject and thought of many other possible experiments. Some of these he actually performed; others he described in detail so that other scientists could try them. He sent letters to London describing his experiments, including those he had done himself and those which he suggested might be done. In the letters he also gave reasons for considering that lightning and electricity were similar in many ways.

Franklin's letters on electricity received great publicity and were translated into French. A well-known French scientist procured a copy but the translation was so bad that he asked a fellow scientist named D'Alibard to revise it.[1] Not only did D'Alibard do so, but, as a result of translating them, he himself became interested in the subject-matter of the letters to such an extent that he decided to try one of the experiments which Franklin had written about, but not performed. This was the experiment of bringing lightning from the clouds to the earth so that its resemblance to electricity could be tested.[2]

In the spring of the year 1752 he engaged an old soldier named Coiffier, who, on leaving the army, had become a carpenter. Coiffier was ordered to make the necessary apparatus and set it up in a sentry box at Marly-la-Ville, a village about fifteen miles from Paris. He made 'an electrical stool' which was simply a board with three wine bottles placed under it to insulate it (glass being a non-conductor of electricity) and obtained an iron rod, about forty feet long and an inch in diameter. He fastened this rod to the stool so that one end pointed up into the air. D'Alibard had told Coiffier, a man on whose intelligence and courage he could depend, to go at once to the sentry box on the approach of the first thunderstorm. He was given a brass wire mounted in a glass bottle for a handle, which insulated it, and told to hold the wire near to the iron rod.

On Wednesday, 10 May 1752, between two and three in the afternoon, Coiffier heard a loud clap of thunder, and rushed to the sentry box. He held the brass wire close to the iron rod. Immediately a bright spark jumped from the rod to the wire with a crackling noise. He then drew a second spark; this was brighter than the first and made more noise. Previously, D'Alibard had told him that if anything unusual ever happened he should send for the priest who would check the observations. So Coiffier sent for him. The priest, on receiving the message, hurried as fast as he could to the sentry box. Some of the parishioners who had heard the crackling sound from the sentry box, on seeing their priest's haste and excitement, started a rumour that Coiffier had been struck by lightning. The rumour spread through the village, and even the hail-storm which followed the thunder did not prevent the flock from 'following its shepherd to watch him administering the last rites to Coiffier'.

34. *The Famous Kite of a Famous Statesman*

But instead of seeing their priest at prayer with a dying man they found that he was holding a wire in his hand and putting one end of it close to the iron rod. Then some of them also saw a bright blue spark, an inch and a half in length, jumping across the gap between the wire and the rod, accompanied by a strong smell of sulphur. Then came another flash which hit the priest on the arm, causing him much pain. His arm had been nearer to the rod than the wire had been. He bared his arm to see what had happened and saw a mark such as might have been made by a blow with the wire on his naked skin. Several people who went near him said that he smelt strongly of sulphur.

The fame of this experiment spread rapidly, and in a few days' time a similar experiment was made in Paris at the request of the King, who saw the sparks with great satisfaction.

Unaware of all the excitement about this public experiment – for news travelled slowly in 1752 – Franklin decided to do the experiment himself by having a long rod erected on top of a large building. But whilst he waited for this to be done, it occurred to him that a common kite could reach higher into the sky than any building. And so he set about making what became the most famous kite in the history of science.[3]

He tied the corners of a large silk handkerchief to a cross made of two strips of cedar and fastened a long iron wire to the upright strip, allowing it to protrude about a foot above the top of the kite. He decided to use a long string of hemp to fly the kite but reasoned that, when the electricity from the 'clouds' was conducted down the wet string and reached the hand, the person who was flying the kite would get a severe shock. So he fastened a silk ribbon to the end of the string next to his hand and suspended a large iron key at the junction of the string and silk. He could then fly the kite in safety, by holding the silk, which is a non-conductor, and keeping under cover so that the silk did not get wet in the rain. He could tell whether any electricity was flowing down the string by putting a knuckle near the key. If sparks passed between his knuckle and the key, and futher, if he experienced a shock, he would know for certain that electricity was flowing down the string.

Franklin, according to a description given by Priestley, 'took the opportunity of the first approaching thunderstorm to take a walk in the fields in which there was a shed convenient for his purpose. But dreading the ridicule which too commonly attends unsuccessful attempts in science, he communicated his intended experiment to nobody but his son who assisted him in raising the kite'.

Franklin's kite

34. The Famous Kite of a Famous Statesman

One day in June, 1752, they took shelter in the doorway of the shed, so that the silk ribbon and key would not get wet, and flew the kite from there. Then, in Priestley's words:

The kite being raised, a considerable time elapsed before there was any appearance of its being electrified. One very promising cloud passed over it without effect; at length, just as he was beginning to despair of his contrivance, he observed signs which showed him that an electric current was passing down the wet hempen string. He immediately presented his knuckle to the key (let the reader judge of the exquisite pleasure he must have felt at that moment), and the discovery was complete. He perceived a very evident electric spark,

or, as Franklin himself wrote later, 'the electric fire streamed out plentifully from the key on the approach of your knuckle'.[4]

The illustration shows Franklin with his son (who was then twenty-three and not a young boy, as is often stated) presenting his knuckle to the key, from which he got a number of sparks. He later charged a Leyden jar by holding its knob to the key. The jar behaved as if it had been charged in the ordinary way. Thus he showed that electricity and lightning were the same.

Franklin was a practical man and decided to apply his discovery that lightning could be brought from the clouds to the earth in the following way. The statement is in his own words:

It has pleased God in His goodness to mankind to discover to them the means of securing their habitations and other buildings from mischief by thunder and lightning. The method is this: provide a small iron rod of such a length that, one end being three or four feet in the moist ground, the other may be six or eight feet above the highest part of the building. To the upper end of the rod fasten about a foot of brass wire the size of a common knitting needle, sharpened to a fine point. A house thus furnished will not be damaged by lightning, it being attracted by the points and passing through the metal (to the earth) without hurting anybody.[5]

Franklin knew that the iron rod was a source of great danger to anyone touching it or coming near it whilst 'lightning' was passing down it. He also knew that anyone who had touched or approached the wet hempen cord which he had used to fly his electric kite would have been in similar danger. This was most tragically first brought home to scientists in the year 1753.

Professor Richmann was then experimenting in St Petersburgh

and had made a piece of apparatus to study the electricity obtained from the clouds. At the approach of a thunderstorm, he went to examine his apparatus and was standing with his head about a foot away from it. His assistant has described how:

all at once a globe of blue fire, as big as his fist, jumped from the apparatus towards the head of the professor. The fire was attended with a report as loud as that of a pistol and parts of the apparatus were thrown about the room; even the door itself was torn off its hinges and thrown into the room. The professor was killed immediately and his left foot was bruised with a blue mark.

From this it seemed, to use the words of the doctor who was called in, that 'the electrical force of the thunder had entered the professor's head and made its way out again at that foot'.[6]

It must be emphasised that all the experiments mentioned so far in this chapter are highly dangerous and must not be repeated. Coiffier, the priest and Franklin were lucky to escape injury.

*   *   *   *   *

From 1753 onwards many lightning conductors (then called Franklin's rods) were erected in America and the practice soon spread to England. Thus the Eddystone Lighthouse was 'protected' by such a rod in 1760, and Franklin's advice on the use of lightning rods became much sought-after. In 1769 he was a leading member of a committee to advise the Dean and Chapter of St Paul's Cathedral in London on the protection of the building from the effects of lightning. In 1772, after an Italian powder magazine had been struck and destroyed by lightning, he was one of a committee appointed to advise on the protection of the British powder magazine at Purfleet.

Some of the members of the committee recommended the use of conductors with round or blunted ends. But Franklin steadfastly urged pointed ones, emphasising that they had been found to be efficient in America. Franklin's advice was taken and pointed ones were fitted. It chanced shortly afterwards that the powder magazine was hit by lightning; but the damage caused was slight and the powder did not explode.

34. *The Famous Kite of a Famous Statesman*

The story so far has dealt with Franklin the scientist; the remaining part of it will deal mostly with his work as a statesman. About the middle of the eighteenth century there were two and a half million people living on the eastern seaboard of North America, most of them being descendants of the early settlers who had left Europe a century or more ago. Those early settlers had left their native lands for a variety of reasons; some to worship as they wished, others to live a life of greater freedom than they could in Europe and others for different reasons again.

They lived in thirteen different settlements or colonies, each of which was, to a great extent, self-governing, but all had one common tie – they were all British colonies and so the people were subjects of the British sovereign. It is now a matter of history that this state of affairs proved most unsatisfactory. In 1776 the colonies decided to become independent of the mother country and so the United States of America came into being. During the preceding few years skirmishes between the colonists and the British troops stationed in America had taken place, but after the Declaration of Independence in 1776 preparations for war were actively begun. Bitter fighting took place until 1783, when England was forced to recognise the independence of her former colonies.

Benjamin Franklin had taken a very active part in these proceedings, and was one of the five colonial statesmen who signed the famous Declaration of the Fourth of July of 1766. This resolved, 'That these United Colonies are, and of right ought to be, free and independent States, that they are absolved from all allegiance to the British Crown and that all political connection between the state of Great Britain is and ought to be dissolved'. Thus the United States of America came into existence.

National feelings and sentiments then, as now, were deeply affected by war. Anything connected with the American rebels, and especially with their leaders, like Franklin, was detestable to many Englishmen. The controversy over the use of pointed or blunt lightning conductors took on a different aspect. Because Franklin recommended pointed conductors those who favoured the use of this type were in danger of being branded as bad patriots. George III led the way and ordered that the pointed rods, having been recommended by a rebel, should be removed from the Government powder magazines and from his own palace and replaced by blunt ones.

But although this was the royal and popular view, the leading scientists did not allow politics to influence them. It is said that the King, not satisfied with changing the rods, tried to compel the President of the Royal Society – the leading scientist of his day – to declare that blunt rods were safer than pointed ones. But the president, Sir John Pringle, refused to do so and replied to the King that 'Duty as well as inclination would always induce him to execute His Majesty's wishes to the utmost of his power, but that he could not reverse the laws and operations of Nature'. Thus, once again, as has happened not infrequently in the history of science, the true scientist refused, even at the request of his king, to be browbeaten into affirming what he believed to be scientifically false.

Franklin was then in France as the representative of the rebel colonies and commented: 'The King's changing his pointed conductors for blunt ones is a matter of small importance to me'. He went further, adding that he wished the King would refuse to use any kind of lightning conductor, for he deserved to be struck by lightning. 'For', he said, 'it is only since he thought himself and his family safe from the thunder of Heaven, that he dared to use his own thunder in destroying his innocent subjects'.[7]

The King's action and Franklin's retort led a rhymester to pen these lines:

> While you, great George, for knowledge hunt,
> And sharp conductors change for blunt,
>   The Nation's out of joint;
> Franklin a wiser course pursues,
> And all your thunder useless views,
>   By keeping to the point.[8]

Before peace was declared a marble bust of Franklin was sculptured in France on which was inscribed this famous line: *Eripuit coele fulmen sceptrumque tyrannis* ('he snatched the lightning from Heaven and the sceptre from tyrants').

A well-known writer, surveying Franklin's life-work, picked out the two incidents mentioned in this story and commented on them in this manner: 'The strong sensation of delight which Franklin experienced as he touched the key of his electric kite might have been equalled, but it was probably not surpassed, when the same hand signed the long-disputed independence of his country.'[9]

35. *Frog Soup and the Electric Battery*

THE MOST famous frog in the history of science belonged to the family of edible frogs. The back legs of these frogs have been regarded as a great delicacy by the French and some southern Europeans for many centuries. The flesh tastes like that of a young chicken or the tender parts of a young rabbit, and is often served fried; but, at the time of this story, soup made from the hind legs was sometimes ordered by physicians for their more delicate patients. It was thought to be very 'strengthening' or 'restorative'. These edible frogs differ slightly from the frogs commonly found in the British Isles, which are not eatable.

About the year 1786 the wife of Professor Galvani, of Bologna University in Italy, was ill, and to hasten her recovery the physician had ordered a soup made of boiled frogs' legs.[1] The professor, like other scientists of his day, experimented in one of the rooms at home to which went many of his pupils for instruction. His wife often sat in the room and watched her husband at work.

One day, so the traditional story relates, Signora Galvani was sitting in this room skinning a number of frogs for her soup; as each was done she put it on a metal dish lying on the table near to her husband's electrical machine. When she had made all the frogs ready she put the skinning knife on the plate and began talking to some of Galvani's pupils who were waiting for him to arrive to begin his experiments. Signora Galvani sat near the table 'keeping her eyes', so the story tells, 'on the delicate morsels', anticipating their fine flavour and the benefit she would get from eating them.

The students were amusing themselves by producing sparks from the electrical machine. Suddenly the Signora saw the frog's legs twitch in the dish as if they were alive. Greatly astonished, she kept her eyes on them for some time. At last she noticed that only those twitched that were touching the knife blade, which was resting on the edge of the metal dish. She also noticed that this twitching seemed to occur only when sparks were being produced by the electrical machine.

She kept her observations private till the return of her husband, who was delighted with the information. He repeated and varied the experiments until finally he made a most important discovery.

This is the traditional account of the way in which Galvani was led to make his series of experiments, and it gives an air of romance to them. But although he, himself, gave a different one there is, nevertheless, some basis for accepting this traditional version. Lucia, his wife, was no stranger to science for she had lived all her life amongst scientists. She was an intelligent woman, being the daughter of a distinguished professor, and after marriage she and her husband shared her father's house which numerous scientists visited. There can be little doubt that if she had noticed the contraction or twitching of the muscles, as related, she would have attached great importance to such a striking occurrence and would have drawn her husband's attention to it at the earliest possible moment.

There is also evidence that she was an invalid about the year 1786, and it is known that she died before Galvani published his account of the discovery.[2]

This relates that he had dissected a frog, leaving its back legs still attached to the spinal cord by the sciatic nerve, and had put them on a table where there happened to be an electrical machine in use. An assistant casually touched the nerve with a dissecting knife and noticed that the muscles appeared to contract repeatedly, 'just as though they had been seized with a violent cramp'. It seemed that this occurred only when a spark was being drawn from the machine and Galvani's attention was drawn to this curious occurrence. Galvani, to use his own words, 'was at once seized with zeal and an intense desire to investigate this occurrence'. He therefore 'touched first one and then the other nerves of the leg whilst at the same moment one of his assistants drew a spark from the machine. The result was the same each time; without exception, sharp contractions seized the muscles of the legs at the exact moment when a spark was drawn from the machine'. It was 'just as though the dissected creature had been struck by tetanus'.

Galvani had previously studied the muscular movement of frogs and had published a paper on the subject in 1772. He, and others, knew before 1786 that the muscles of animals twitched or became convulsed on being put in direct contract with a Leyden jar or an electrical machine. It is, therefore, reasonable to accept his account, for he may well have been getting ready to do an experiment in which he required frog's legs and an electrical machine.

Each of the versions, however, agree in this respect, that a chance observation first drew attention to the twitching of the muscles whilst the machine was giving off sparks. One credits the wife, the other the assistant, with noticing the twitching.

As a result of this chance observation Galvani made a series of experiments on frogs' legs. The one most frequently mentioned is that in which he hung a frog (or frogs) on the iron railings surrounding the balcony of their house.

All accounts of this experiment agree that he wished to study the effect of lightning, that is of atmospheric electricity, on the muscles. The story, as often told, is that one fine day, when the sky was calm and the wind was blowing gently, he happened to see the frog's legs twitching when the wind blew them against the railings. This greatly amazed him for the convulsion could not have been caused by atmospheric lightning. This chance observation started him on another long series of experiments.

Galvani, however, again gave a different account. He dissected a frog, he wrote, leaving the pair of legs attached to a stump of the backbone by the sciatic nerves, and put a brass hook through the spinal cord, thus ensuring that it touched the nerves. Then he hung the legs on the railings. He noticed that the muscles contracted or twitched occasionally during fine weather as well as during thunderstorms, and so kept close watch on them for many fine days. But only rarely did he see a twitching. At last, 'tired of fruitless waiting' he decided to push the brass hook against the iron railing to see whether the muscles would then twitch. They did. He repeated this time and time again, and there were frequent twitchings. None of them had any connection with atmospheric electricity for the weather was fine.

Thus, according to his account, he discovered this new kind of twitching not by accident, as is sometimes stated, but as the result of a deliberate act brought on by an understandable impatience.

This incident on the iron railings, whether he observed it by chance or not, sent him indoors with the legs to make a simple test. He laid them on an iron plate and pressed the brass hook against the plate. Each time he did this, to use his words, he 'beheld the same twitchings'.

Galvani, after many other experiments, then tried to explain why the muscles twitched. He knew, as has already been mentioned, that muscles of animals twitch when they are put in direct contact with the electric machine. Electricity evidently brings on the twitching. This new twitching, however, took place without the application of electricity externally. Hence he thought that it must be caused by 'indwelling animal electricity'. This animal electricity, he stated, flowed through the nerves to the muscles, but it did so only when two different metals formed part of the circuit. This is illustrated in the drawing which shows Galvani holding a bent piece of *iron* which touches the *brass* ring around the spinal cord at the top of the legs, and also the frog's toes.

His investigations and his theory of the existence of animal electricity were no sooner known to the scientific world than they excited general interest. His experiments were varied in many ways, a prominent worker being his countryman Alessandro Volta, a professor from Padua.

35. *Frog Soup and the Electric Battery*

The frog's legs twitch

At first Volta accepted Galvani's theory, but after a series of experiments rejected it, for he showed that the nerves and muscles of frogs had nothing to do with the generation of electricity – animal electricity did not exist. Instead, as he proved, the electric current was produced by the contact of two different metals. So thoroughly did he prove this that he was able to invent what is now known as a 'voltaic pile'. This pile was made up of a plate or disc of silver on top of which was a disc of zinc of the same size or shape. Then came a layer or disc of flannel which had been moistened previously in a salt solution. Next came another pair of silver and zinc discs, followed by another flannel disc and so on, until a pile of twelve metallic discs had been formed.

When Volta touched the top disc with one hand and the bottom one with the other he felt an electric shock; when he connected the top and bottom discs by a wire electricity passed along it in a continuous flow.

The pile gave a chemical method of obtaining current electricity, and one of its main advantages was that the supply was continuous. Its use greatly enlarged the kind of experiments which scientists could make. For example, within a few years of its discovery, Sir Humphrey Davy had used it to obtain the metal sodium for the very first time. Other chemical discoveries were made by its use whilst the study of electricity was greatly advanced. This pile was the forerunner of all the nineteenth-century batteries and cells and its chemical principles are still the basis on which modern ones are made.

It is interesting to recall that these numerous advances in science have a remarkable origin. There is the chance observation made in Galvani's house in 1786; possibly there is another chance observation of the frog on the iron railings; and there certainly is the mistaken theory of animal electricity.

Unhappily Galvani got little lasting pleasure from his labours. His devoted wife Lucia died even before he could publish an account of his work; and he had many other worries. For the French Revolution upset the old order of government and a republic had been set up in his country. Galvani, it is said, would not promise obedience to the new rulers and not only was deprived of his professorship and the income attached to it but had to leave his home and find residence elsewhere. He went into quiet retirement to his brother's house, where he fell ill and died in 1798.

It is said that the republican government, in consideration of his great scientific fame, determined to reinstate him in his professorial chair at the University of Bologna. But they were too late.[3]

36. The Rival Claims of Two Inventors

COAL MINING has always been a dangerous occupation, not only because of the likelihood of injury from falling rock, but also because of the presence of the gas known as fire-damp. This gas, whose chemical name is methane (with the formula CH_4), is present in all coal mines. It escapes with great force from cracks or gaps in the coal layer making a noise like that of a strong wind. For this reason the cracks were known as 'blowers' at the date of this story.

Fire-damp mixed with from four to twelve times its own volume of air explodes when a light is applied to it, the most explosive mixture being composed of one part of gas to seven or eight parts of air. The gas mixed with more than twelve times its own volume burns quietly with a pale blue flame. In these respects it is very similar to the coal gas from the gas works.

Miners, who had to have some light in the dark mines, used candles for many years but they could do so only in those parts of the mines where the proportion of gas and air gave a non-explosive mixture. The use of naked lights in coal mines, has, therefore, always been dangerous.

An interesting method of providing illumination was used in a few mines near the sea. The scales of some fish possess the strange property of glowing in the dark, so miners used to fix them on to a wooden board and take that into the mine. The glow from the scales gave a faint glimmer of light.

About the year 1740 a Whitehaven mining engineer named Spedding invented an illuminating machine called a steel mill which produced sparks. Thus it could be used in places where the mixture of gas and air was not highly explosive; that is, was one which was exploded by a naked candle light but not by a spark. It consisted mainly of a piece of flint fixed against the jagged edge of a steel wheel. Sparks were produced when the wheel was turned rapidly round by hand so that the jagged edge rubbed against the flint. The same method of producing sparks is used in a modern cigarette lighter.

In 1813 the Society for the Prevention of Accidents in Coal Mines was formed and many influential people in the north of England became members of it. One such member was the Rev. John Hodgson, Vicar of Hewarth, near Newcastle. He had a good knowledge of mining and was particularly interested in this Society, especially after a terrible accident at a pit in his neighbourhood when more than ninety men were killed. In the year 1815 Hodgson met the eminent scientist, Sir Humphry Davy, who was travelling in the north, and they discussed safety devices in mines.

Sir Humphry became interested in the problem and in the course of a few months had invented his safety lamp, which is illustrated on p. 91. It consisted of a lighted wick which was surrounded by an iron gauze through the holes of which the air necessary for burning could pass in and the products of combustion pass out. But the flame did not pass through the holes and so the surrounding explosive mixture of fire-damp and air was not set alight. In later models, a glass cylinder was substituted for the lowest part of the gauze.

Sir Humphry sent his first lamp to Mr Hodgson and asked him to try it out. The first trials were made at the mouth of an iron pipe which discharged fire-damp from the mines. The lamp burned brightly but safely. Mr Hodgson then tried it in the well-ventilated parts of the mines and again found that it burned with safety. This encouraged him to make a more serious trial, or, as his biographer puts it:[1]

to remove the most distant idea of doubt respecting the safety afforded by the lamp, by taking it himself to a part of the mine which could not be ventilated without the greatest difficulty, – a place which was considered quite unsafe for men to work in by the light of a candle. Here a man was working, hewing the pillars of coal by the light of a steel mill.

His account continues:

No notice had been given to the man of what was about to take place. He was alone, in an atmosphere of great danger, in the midst of life and death,

36. *The Rival Claims of Two Inventors*

when he saw a light approaching – apparently a candle burning openly –
the effect of which he knew would be instant destruction to him and its
bearer. His command was instantly, 'Put out that candle'. It came nearer
and nearer: no regard was paid to his cries, which then became of the most
terrible kind, mingled with awful cursings of his comrade – for such he took

Put out that candle

Mr Hodgson to be – who was tempting death in so rash and certain a way.
still not one word was said in reply. The light continued to approach, and
then oaths were turned into prayers that his request might be granted, until
there stood before him, silently exulting in his heart, a grave and thoughtful
man, a man whom he well knew and respected, – who, four years before,
after a terrible pit explosion, had buried in one common grave ninety-one
of his fellow-workmen, – holding up in his sight with a gentle smile the
triumph of science, the future safeguard of the pitmen.

A few months afterwards Mr Hodgson reported as follows to
Sir Humphry:

The conversations of the common colliers respecting the lamp are very
amusing and interesting. They have not yet ceased to wonder at its magic-
like properties, and seem divided in their opinion whether to regard it as
something supernatural or an instrument subject to the common laws of
causes and effects.

Davy read a paper describing this lamp before the Royal
Society on 9 November 1815, and exhibited a model of the
safety lamp to the members.

Some years before this date, a humble millwright named George Stephenson had been making experiments on fire-damp, and, before any public statement about the Davy Lamp was made, had tried out his first and second models, on 21 October and 4 November 1815. On 30 November he had constructed and tested his third and final model. An account of some of his trials reads:

He, with two others, Moodie and Wood, went to one of the galleries in the pit which contained the most explosive gas which was making a hissing noise as it issued from the coal face. They erected some deal boarding round that part of the gallery into which the gas was escaping; the air was thus made more foul for the purpose of the experiment.

After waiting for an hour, Moodie, whose practical experiences of fire-damp in pits was greater than that of either Stephenson or Wood, was requested to go into the place which had thus been made foul; having done so, he returned, and told them that the smell of the air was such, that if a lighted candle was introduced, an explosion must inevitably take place. He cautioned Stephenson as to the danger, both to themselves and to the pit, if the gas took fire. But Stephenson declared his confidence in the safety of his lamp, and, having lit the wick, he boldly proceeded with it towards the explosive air. The others, more timid and doubtful, hung back when they came within hearing of the blower.

Apprehensive of the danger, they retired into a safe place, out of sight of the lamp, which gradually disappeared with its bearer into the recess of the mine. Advancing to the place of danger, and entering within the foul air, his lighted lamp in hand, Stephenson held it firmly out, into the full current of the blower, and within a few inches of its mouth. Thus exposed, the flame of the lamp at first increased, then flickered and then went out, but there was no explosion.

This and other experimental trials showed that he must modify the lamp, and on 30 November 1815, he produced his third and final model. The illustration shows the Davy Lamp side by side with Stephenson's. The flame in Stephenson's lamp is surrounded by a glass cylinder, which is covered by an iron plate with holes in it.[2]

A bitter controversy was waged about the respective merits of the two lamps and as to which was discovered first. It would seem that Davy tackled the problem in the laboratory from a chemical standpoint, studying the explosive nature of various mixtures of gas and air. He also found that the flame of fire-damp does not pass through the holes in wire gauze. Stephenson

36. The Rival Claims of Two Inventors

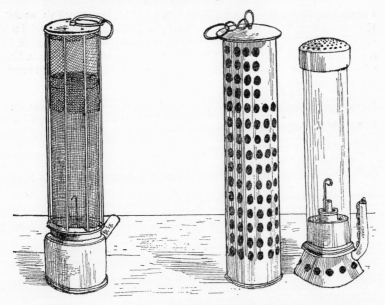

Davy's and Stephenson's lamps

tackled the problem more mechanically by trying different kinds of lamps in the mine. He was the first of the two to observe that explosions do not pass through tubes.

Sir Humphry Davy carried off the larger share of the praise which attached to the discovery; he was the scientific prodigy of his day, the most brilliant of lecturers and the most popular of philosophers. Stephenson, on the other hand, was then but an engine-wright of the colliery, scarce raised above the manual-labour class, and this encouraged one critic to comment thus:

It will hereafter be scarcely believed that an invention so eminently scientific, and which could never have been derived but from the sterling treasury of science, should have been claimed on behalf of an engine-wright of the name of Stephenson – a person not even possessing a knowledge of the elements of chemistry.[3]

How mistaken this writer was, for a brilliant future awaited Stephenson; for he was the Stephenson who became the world's most famous railway engineer.

Davy laid claim to his priority thus:

I never heard a word of George Stephenson and his lamp until six weeks after my principle of security had been published and the general impression of the scientific men in London, which is confirmed by what I heard at Newcastle, is that Stephenson had some loose idea floating in his mind which he had unsuccessfully attempted to put into practice till after my labours were made known.

He added that there was no similarity between 'Stephenson's glass exploding machine' and 'his own metallic tissue which permits light and air to pass through it but not a flame'.

The leading chemists and natural philosophers of the country, with the President of the Royal Society, held an enquiry in 1817 and concluded that Sir Humphry Davy discovered the lamp 'independently of all others'.

Thereupon a meeting of owners of coal mines opened a subscription list and gave two thousand pounds to Sir Humphry and 'a purse of one hundred guineas to George Stephenson in consideration of what he had done in the same direction'.[4] But Stephenson's friends also called a meeting and passed this resolution:

That it is the opinion of this meeting that Mr George Stephenson, having discovered the fact that explosion caused by hydrogen gas will not pass through tubes and apertures of small dimensions and having been the first to apply that principle in the construction of a safety lamp, is entitled to a public reward.

A sum of a thousand pounds was collected, and the 'humble colliers' bought Stephenson a silver watch, a gift which gave him much pleasure.

A happy way of closing this bitter controversy was taken by Robert Stephenson, the famous son of a famous father, when some forty years later, he was asked for his opinion. He said:

I am not exactly the person to give an unbiased opinion; but as you ask me frankly, I will as frankly say, that if George Stephenson had never lived, Sir Humphry Davy could, and most probably would, have invented the safety lamp; but again, if Sir Humphry Davy had never lived, George Stephenson certainly would have invented the safety lamp, as I believe he did, independent of all that Sir Humphry Davy had ever done in the matter.[5]

37. Soldiers Break Step on Suspension Bridges

BEFORE A body of soldiers marches across a supension bridge the officer in charge must give the order to 'break step'. This is necessary because of the way in which this type of bridge is constructed.

Suspension bridges have been built of iron since early in the nineteenth century. In their construction a strong pillar or tower is first built at each end of the gap which is to be bridged.

A suspension bridge

A long, strong, iron chain or cable is then hauled into position; that is, one end is laid on the ground at one side of the gap and the other end is first passed over the top of one pillar and then across the gap and over the facing pillar on the other side, in such a way that the chain passes over the tops of both pillars. Each end is fastened to something firmly bedded in the ground, such as a piece of natural rock or a large block of iron or concrete specially buried in the ground for the purpose. Another chain is suspended in this way across the gap over the pillars. These chains hang in graceful curves over the gap and to them is fastened, by iron rods or chains, the framework which carries the main flooring or deck of the bridge – the part which traffic will use. The great advantage of building a suspension bridge is that a smaller amount of material is required for its construction than for most other types of bridge.

Most readers will have seen an acrobat bouncing up and down on a tightly sprung canvas which acts like a large, springy mattress and is known as a trampolin. At first he steps up and down on it in an irregular way, but before long he is bouncing high into the air. At the end of his performance he contrives, by the way he treads on it, to come to rest, standing on the trampolin.

Under the movements of the performer the canvas is, of course, made to move up and down, that is, it is put into a state of vibration. When vibrating in a natural way it always completes the up and down movements in the *same* time – the farther it moves up and down the faster it moves, and conversely, the less it moves up and down, the slower it moves. The rate of this regular movement may be termed the natural vibration-time of the trampolin.

The performer first sets the canvas in motion. He steps or jumps on the canvas at varying rates until he feels that his movements are beginning to fit in with those of the trampolin. He and the trampolin are then moving at its natural vibration time. Every push which his feet give to the trampolin in its downward movement helps it to go still further downwards and so throws him higher in the air on its rebound.

When he wishes to stop he begins to tread out of phase, and the force of impact of his feet acts against the trampolin's regular movement up and down, which quickly decreases so that in a short time the canvas comes to rest.

Strange though it may seem, a circus performer treading in phase with a trampolin may be compared with a body of soldiers marching in step across a suspension bridge, for such a bridge may be set into vibration in its 'natural vibration time' by a series of uniform steps so that it ultimately swings to a dangerous extent. This has actually happened on two recorded occasions when soldiers have been *marching in step* over such a bridge, with disastrous results.

37. *Soldiers Break Step on Suspension Bridges*

The first of these occurred on 11 April 1831, when men of the 60th Rifle Corps were marching back to their quarters after a morning spent on field exercises. Their route lay across a chain suspension bridge, fifty yards long, which spanned the River Irwin, connecting Pendleton with Broughton, near Manchester. (This river is tidal and flows into the Mersey.) The bridge was privately owned, and by a very strange coincidence, the officer in charge of the troop of sixty-eight men was Lieutenant Fitzgerald, the son of the owner of the bridge.[1]

The officer was leading his men, four abreast, and had arrived near the centre of the bridge when a dreadful crash was heard 'resembling a continuous discharge of musketry'. In a moment one side of the bridge had dropped into the river, dragging with it the main pillar. The scene which followed can be well imagined.[2] All the soldiers on the broken part of the bridge were thrown either into the river or among the chains, their rifles and equipment being scattered in all directions. Fortunately the tide was out, so that the river was only a few feet deep: there can be little doubt that otherwise many of them would have been drowned. Fortunately also, the injuries were less than they might well have been; one man had his leg broken and another his arm; six were seriously injured, a few of them being disabled for life.[3]

A newspaper commented:

It has been stated by some scientific men, and we fully concur in the opinion, that the peculiar manner in which the soldiers marched whilst on the bridge had no small share in causing the accident. Before they reached the bridge, we are told, they were marching at ease but when they heard the sounds of their own footsteps upon it, one or two of them began to whistle a martial tune and they all at once, as if under a command from their officer, commenced marching together in step. The bridge was only a few years old but it had been crossed daily by carts and waggons; indeed, the rifle party had themselves passed over the bridge the same morning, walking across it in an easy manner without using the military march; several waggons had also crossed it the same day.

Another newspaper stated:

There is no doubt that the immediate cause was the powerful vibration communicated to the bridge by the measured and uniform step of the soldiers. If the same or a much larger number of persons had passed over in a crowd, and without observing any regular step, in all probability the accident would not have happened, because the tread of one person would have counteracted the vibration arising from that of the other. But the soldiers, all stepping at the same time, and at regular intervals, communicated a powerful vibration to the bridge which went on increasing with every successive step and which, causing the weight of the bridge to act with successive jerks on the stay chains, had a more powerful effect upon them than a dead weight of a much larger amount would have had.[4]

The accident happened early in the history of iron suspension bridges in Britain, for the Menai suspension bridge, one of the first and best known of this type of bridge, had been erected only about 1820, that is, some ten years earlier. Most people had no acquaintance with these bridges and began to wonder whether this long bridge at Menai, with a span of 570 feet, was safe. A Manchester paper had this to say:

Perhaps the accident, alarming and injurious as it has been, may have the effect of preventing some more dreadful catastrophe in other quarters. From what has happened on this occasion we should greatly doubt the stability of the great Menai Bridge if a thousand men were to be marched across it in close column and keeping a regular step. From its great length, the vibrations would be tremendous before the head of the column had reached the other side and some terrific calamity would be very likely to happen. If any considerable number of troops should be marched across that bridge we hope the commanding officer will take the precaution of dismissing his men from their ranks before they attempt to cross. Indeed, the precaution should be observed by troops crossing all chain bridges, however small they may be.[5]

Nineteen years later a similar accident occurred. A suspension bridge had been built, about twelve years previously, across the River Maine at Angers in the old Province of Anjou of France. In 1849 it had been inspected and repaired at a cost of 36,000 francs. A newspaper account runs as follows:

On the 16th April, 1850, at eleven o'clock in the morning two battalions of the line and a squadron of Hussars had crossed over this bridge without any accident. The last of the horses had scarcely crossed when the head of the column of the 3rd Battalion of the 11th Light Infantry appeared on the other side. Repeated orders were given to the troops to break into sections, as is usually done, but the rain was falling heavily at the time and the order was disregarded. So the battalion marched over the bridge in close column, marching in step. The head of the column had reached the opposite side, indeed, the pioneers, the drummers and a part of the band were off the bridge, when a horrible crash was heard. The suspension chains on one side had given way under the measured tread of the soldiers. The soldiers on the bridge, feeling the movement of the flooring, rushed to the other side, when the chains there also gave way. The whole flooring of the bridge then fell into the river and with it all the soldiers who were on it. From one

Catastrophe of the bridge at Angers
(Based on an illustration from a contemporary newspaper)

bank to the other the river was completely blocked with the soldiers struggling to reach the shore. By this terrible catastrophe, a captain, a lieutenant, three sub-lieutenants and two hundred and twenty-one rank and file lost their lives; and it is supposed that there were a considerable number of women and townspeople accompanying this regiment who also perished.

The bridge is the ordinary passage for the troops, and the most direct route to the castle. To describe the frightful spectacle and the cries of despair which were raised is impossible. The whole town rushed to the spot to give assistance. In spite of the storm which was raging, all the boats that could be got at were launched to pick up the soldiers in the river and a great number who were clinging to the parapets of the bridge, or who were kept afloat by their knapsacks, were got out. The greater number were, however, found to be wounded by the bayonets or by the fragments of the bridge falling on them.[6]

Another newspaper added:

From one bank to the other the river was completely blocked with the soldiers struggling to reach the shore. If the weather had been calm, the greater number of them would in all probability have been saved. The wind, however, blew a perfect hurricane and the waves were very rough. Masses of the men might be seen clinging to each other, the waves every moment washing away some of them until only one remained. Beams of wood, planks and every article which could be laid hold of, were launched to enable the men to keep themselves afloat until further assistance could arrive.[7]

An ugly rumour was started that this regiment was being sent, as a punishment, to Africa and that the men in consequence were disobedient and deliberately refused to obey the order when they were commanded to break step. This rumour was, however, officially denied.

38. The Plimsoll Mark

BRITISH SHIPS, with a few exceptions such as yachts and fishing vessels, have lines painted on their sides to indicate the limits to which the ship may be loaded without it being in danger of sinking. The position of the lines depends on the temperature and degree of saltiness of the water in which the ship will sail. The set of lines is called the Plimsoll Mark after Samuel Plimsoll (Liberal Member of Parliament for Derby between the years 1868 and 1880) who played a historical and sensational part in the passing of the Act which made this compulsory.[1]

Merchant shipping in the mid-nineteenth century came under severe criticism from a Newcastle shipowner named James Hall. He was deeply shocked at the condition of some of the ships which put out to sea, many of which, so he said, besides being unseaworthy, were overloaded or had insufficient or inefficient apparatus with which to work the ship. As a result of these conditions many ships were lost at sea, often with heavy loss of human life. These 'coffin ships', as they became known, were often over-insured so that the unscrupulous shipowner did not lose money – indeed, on the contrary, he often made a handsome profit when his ship went down.

38. *The Plimsoll Mark*

The law was then not adequate to put an end to these appalling conditions so Hall began a crusade to have it amended. In this he was only partially successful, for an Act passed in 1871 to deal with the situation did not make overloading illegal, as Hall had hoped.

While the Bill was passing through Parliament, Hall met Samuel Plimsoll and the latter learned, probably for the first time, exactly how appalling some of the conditions were in the mercantile marine. Hall's proposals were of just the type to appeal to the humanitarian nature of the newly elected member for Derby, who soon took the conduct of the campaign largely out of Hall's hands. Not only did he vigorously advocate most of the reforms proposed by Hall but he adopted them as if he himself had been the author of them. So thoroughly did he throw himself into the campaign that before long it was his name, and not Hall's, that became associated with the proposed reforms in the minds of most people.

The full force of Plimsoll's attack was exerted in 1873 when his now famous book *Our Seamen – An Appeal*[2] appeared. In it he wrote:

There are many hundreds of lives lost annually by shipwreck; and as to the far greater part of them, they are lost from causes which are easily preventable. A great number of ships are regularly sent to sea in such a rotten and otherwise ill-provided state, that they can only reach their destination through fine weather, and a large number of them are so overloaded that it is nearly impossible for them to reach their destination if it is at all rough.

The book was warmly welcomed by many newspapers, and most of the reforms suggested in it gained such a favourable public opinion, that within two months of its publication he confidently moved in the House of Commons that a Royal Commission of Inquiry be set up to investigate and report on many of the matters mentioned in his book. Only a brief extract is necessary to show how convincingly he presented his case. 'I may tell you', Plimsoll said,

'why I feel so strongly on behalf of the seamen. If the lives of nearly a thousand of our ministers of religion, of our doctors, or of our public men were sacrificed every year, to what a Government official calls a homicidal system of most culpable neglect, what would you say? All England would ring with indignation at the outrage. Yet I venture to say that any thousand of what are called the working classes are as worthy of respect and affection as any of these.'[3]

Unhappily Plimsoll allowed his enthusiasm to run away with him, for, although there were many whom he rightly termed ship knackers and wrecking shipowners, there were also many others who were good employers. When deciding whom to criticise he made little effort, if any, to verify the information which came his way. Had he been more careful he would not have accused two or three fellow members of the House of Commons who were also shipowners. But this he did. These members, he stated, had been guilty of practices by which they had made large fortunes, and, having got into Parliament, did what they could to obstruct and prevent legislation on this subject.

This was a most serious allegation and it is not surprising that one of the members took legal action against Plimsoll for criminal libel – a most serious offence. This particular member stated in court that he had never lost a ship by foundering, except by stranding or collision; that he had never lost a ship from stress of weather alone, with one exception; and that he had never lost the life of a single seamen.[4]

After many delays the Court of Common Pleas gave its verdict that Plimsoll was to blame because he had been hasty in making statements upon insufficient evidence; but, the court added, the case was not a suitable one to make Plimsoll subject to criminal law. Plimsoll, however, had to pay his own costs, which amounted to a considerable sum.[5]

The report which the Royal Commission presented gave Plimsoll practically no support in his campaign against overloading. However, he persevered and at last, in 1875, a Bill was brought in to amend the Shipping Acts. The first provision of the Bill enacted that all ships, other than those already under survey by Lloyds or by the Association in Liverpool, should be surveyed before they left their respective ports. The next provisions enacted that there should be a maximum load-line, below which no vessel should be put into water.

The Bill did not have an easy passage and matters came to a

head when Disraeli, the Prime Minister, announced that the Government intended to withdraw the bill. Then came a scene such as is rarely witnessed in Parliament, for Plimsoll completely lost control of himself. In a state of great agitation he stepped forward to move the adjournment of the House. He stated that he would later ask for information from the President of the Board of Trade as to the loss of certain vessels in 1874 and whether they were owned by Mr Bates, one of the members for Plymouth. He added that he would also have similar questions with respect to some of the Liberal members of the House. Then came a startling announcement, 'I am determined', he said, 'to unmask the villains who send these people to their death'.[6]

Plimsoll stood in front of the chair gesticulating violently, stamping his foot, and shaking his clenched fist at the Treasury bench. Such conduct in the House could not, of course, be overlooked; especially the accusation that fellow members had been guilty of villainous conduct. The Prime Minister moved that he should be reprimanded, but his friends spoke in his favour and the Speaker ordered him to appear in his place that day week.

A week later Plimsoll, in 'a crowded and fussy House', expressed his deep regret; and the House accepted his apology.[7] As it turned out the incident did his cause no harm; on the contrary, it attracted so much public attention that the Government was forced, by newspaper comment and public opinion generally, to hurry the Bill through Parliament; and in August 1875 the Bill passed through the House of Commons and became the Merchant Shipping Act, 1876.

39. The Chance Discovery of X-Rays

IN THE later part of the nineteenth century many scientists were studying the remarkable effects produced when electricity is discharged in a partial vacuum and were greatly helped in their work by the invention of Crookes's tube in 1879. This is a long cylindrical tube of glass containing two terminals. One is connected through an induction coil to the positive pole of a battery and is called the anode. The other is similarly connected to the negative pole and is called the cathode. Practically all the air has been pumped out of the tube by a vacuum pump which has been connected to a small outlet tube which is then sealed off.

When the current is switched on the walls of the tube become a ghostly, apple-green, shimmering colour, or, as scientists say, they become fluorescent. From their observations Sir William Crookes and others deduced that this fluorescence was caused by rays which come from the cathode and strike the inner walls of the tube.

A few years later, Professor Lenard discovered that these cathode rays, as they are called, whilst they were stopped by even a thin wall of glass, would pass through aluminium foil; and so he devised an improved kind of tube which had a window of aluminium inserted in its wall. Lenard found that the cathode rays passed out of the tube and into the air where they could be detected for a very short distance only.

A few other substances besides glass fluoresce when cathode rays fall on them, one being barium platino-cyanide; and towards the end of the century many scientists were using a screen made of a sheet of paper or cardboard coated with tiny crystals of this substance in their experiments on cathode rays.

One day towards the end of the year 1895 Professor Röntgen of Wurtzberg in Bavaria was experimenting with an improved Crookes's tube. He had darkened the laboratory by drawing the blinds and had covered the tube with a shield of black cardboard through which no light, however intense, could pass. It was thoroughly dark in the laboratory, therefore, when he switched on the coil. He then happened to glance round and saw that one of the fluorescent screens, which was standing on the table a few feet away, was glowing brightly. This sight puzzled him, for the tube was thoroughly blacked out and there were no signs of cathode rays escaping from it. Yet it seemed as if some kind of rays were coming, in a direct line, from the tube to the screen; for, as he soon proved, they could not have come from anywhere else. He moved the screen nearer, and found that it continued to glow if he kept it pointing in the same direction.

Gradually he became convinced that a new kind of ray was being given out by the fluorescent tube; a ray which could pass through thick black paper. Perhaps, he reflected, it could pass through other things. He put a piece of wood directly between the tube and the screen. The screen glowed – the rays were passing through the wood. He replaced it with a piece of cloth, again the screen glowed – the rays were passing through the cloth. He then interposed a small piece of metal; but the metal cast a shadow of its own shape on the screen – evidently the mysterious rays were not passing through the metal.

The bones are depicted

Then he had a most brilliant yet simple idea. Rays of ordinary light, he reflected, affect a photographic plate; perhaps these mysterious rays would do so also. To test this idea he put a photographic plate in the path which the rays would take and persuaded his wife to place her hand between the tube and the plate. He switched on the coil. When the plate was developed he and his wife saw the bones appearing distinctly with the flesh dimly outlined around them. This was the first time a photograph of the skeleton of any living person had been taken and it must have been a staggering sight for the lady to see this photograph of part of her own skeleton.

*   *   *   *   *

Another version of this wonderful discovery relates that Röntgen had been reading a book and had used an iron key as a bookmarker; he had then put the book down on his laboratory bench, on top of a photographic plate contained in its wooden holder. Later, whilst experimenting with Crookes's tube, he went out of the laboratory for a short time, placing the fluorescing tube on top of the book. On his return he continued his experiments. Some days afterwards, he used this photographic plate to take a photograph of an outdoor scene but when the plate was developed to his great surprise he saw an impression of the key on the negative. This led him to think that the Crookes's tube was giving out a mysterious kind of ray. There is no need to try to verify or disprove this far-fetched version, for few people now believe it.

*   *   *   *   *

Röntgen called the rays X-rays, because little was known about them and mathematicians always use the letter X to denote an unknown quantity. Later, an attempt was made to give them the name of 'Röntgen rays' which would be more appropriate and would honour the discoverer. But the name was not used for long, in this country at any rate. As the editor of a scientific journal has written Professor Röntgen did not have the good fortune of having a name which sounded pleasant when spoken.[2] (Its pronounciation is '*runtyen*', with the 'u' as in ruck, and the 'e' as in 'peck'.)

As with many other chance observations, it seems almost unbelievable that no one before Röntgen had observed what he had just seen, particularly because many astute scientists had been doing experiments with Crookes's tube for fifteen years or more before 1895.

After Röntgen's publication of the details of his discovery Sir William Crookes realised how near he himself had been to making the same discovery. Thus, as Lord Rayleigh, another brilliant physicist, wrote:

It was a source of great annoyance to Crookes that he missed the discovery of the X-rays. According to the account he gave in my hearing, he had definitely found previously unopened boxes of plates in his laboratory to be fogged for no assignable reason, and, acting I suppose in accordance with the usual human instinct of blaming someone else when things go wrong, he complained to the makers who naturally had no satisfactory explanation to offer. I believe that it was only after Röntgen's discovery that he connected this with the use of highly exhausted vacuum tubes in the neighbourhood.[3]

Röntgen first described his discovery to the Physico-Medical Society of Wurzburg in December, 1896, and shortly afterwards particulars were released to the press. The discovery created a great sensation in many countries. Early in January of the following year, a celebrated British professor of physics described the discovery in a journal which was read by most educated people. He began by stating that a very singular scientific discovery had just been made by Professor Röntgen of Wurzburg in Bavaria, who had succeeded in finding a means of photographing an object of metal which was completely enclosed in a wooden box with more ease than if it had been in a glass one. Röntgen, he added, could also photograph a man's skeleton through the skin, flesh and clothes, which under these rays were photographically transparent, while the bones were opaque, like metal.

'The discovery', he continued,

adds one more to the marvels of science. To photograph in total darkness seems inexplicable, but that we should be able to photograph through walls of wood, or through solid opaque bodies, is little short of a miracle. We shall now be able to realise Dickens's fancy when he made Scrooge perceive through Marley's body the two brass buttons on the back of his coat. We shall now be able to discover photographically the position of a bullet in a man's body. Even stone walls will not a prison make to the revelations of the camera.'[4]

Lord Rayleigh also described this announcement when he wrote many years later as follows: 'Röntgen's discovery was followed by a greater outburst of enthusiasm than any other

experimental discovery before or since. Most physical laboratories had the means of taking X-ray photographs of hands and
this was tried on all sides.' For example, almost immediately
after hearing about the discovery, Professor J. J. Thomson gave
a lecture at the Cavendish Laboratory during the course of which
a photograph of the hands of one of the ladies present was taken,
developed and shown during the lecture.[5]

It is not surprising, therefore, that the 'man in the street' was
led to think that Röntgen had invented a kind of camera which
would produce a photograph showing the bones of the body, or
that some newspapers called the discovery 'a revolution in
photography'. Indeed, the editor of a scientific journal, commenting on this aspect, wrote: 'There are very few persons
who would care to sit for a portrait which would show "only
the bones and rings on the fingers".'[6]

Some people became alarmed at the thought that the new
invention would enable, for example, a street photographer to
take intimate photographs 'which would be an insult to de-
cency'. Indeed it is said that an enterprising firm in London not
only advertised underwear guaranteed to be X-ray proof but
actually made a small fortune out of their sales of it.

Punch produced these lines:

> O, Röntgen, then the news is true
> And not a trick of idle rumour,
> That bids us each beware of you
> And your grim and graveyard humour.
> We do not want, like Dr Swift
> To take our flesh off and pose in
> Our bones, or show each little rift
> And joint for you to poke your nose in.

> We only crave to contemplate
> Each other's usual full-dress photo,
> Your worse than 'altogether' state
> Of portraiture we bar in toto.
> The fondest swain would scarcely prize
> A picture of his lady's framework;
> To gaze on this with yearning eyes
> Would probably be voted tame work.[7]

40. *The Discovery of Radioactivity*

Meanwhile, serious students had begun to appreciate the tremendous boon which these rays might confer on mankind. Indeed the importance of X-rays in surgery was at once realised by doctors; and it is perhaps worthy of note that Röntgen's first announcement of his discovery of the rays was read before a Medical Society – that of Würzburg. Thus 'surgery was the first art to be closely identified with X-rays. On January 20th, 1896, a doctor in Berlin detected a glass splinter in a finger; on February 7 a doctor at Liverpool X-rayed a bullet in a boy's head; in April a professor at Manchester took X-ray photographs through the head of a woman who had been shot'.

Years later, Sir J. J. Thomson summarised the value of X-rays to surgery in these words. 'Few have done more to relieve human suffering than Röntgen and those who by developing the application of X-rays to surgery have supplied the surgeon with the most powerful means of diagnosis.'

Doctors have other uses for X-rays, for example, to destroy cancer cells and such diseases as ringworm. Industry has also found them of value, especially in metallurgy, where they can detect flaws and cracks in iron structures which have been cast or moulded.

40. *The Discovery of Radioactivity*

A FEW months after Röntgen's fortunate observation had led him to make his important discovery another scientist, after considering the production of X-rays, made an experiment which had an unexpected result. This unexpected result led to the discovery of radioactivity.

When X-rays are being produced the glass wall of the Crookes's tube, as mentioned on page 102, glows with a greenish shimmer where the cathode rays fall on it. But the glow stops immediately the cathode rays are cut off. The glowing part of the tube is said to be fluorescent.

Fluorescence is not uncommon. It is produced when sunlight falls on a few substances, which then glow with a bluish colour. They lose their glow, however, immediately they are put in the dark. A few other substances glow in a similar way when exposed to sunlight but, unlike fluorescent substances, they continue to glow for a short time after being put in a dark place. These are said to be phosphorescent substances. Phosphorescence and fluorescence, however, are very similar in many ways.

E. Becquerel a well-known French scientist, and his son Henri, during the later part of the nineteenth century specialised in the study of substances containing the then rare metal called uranium. The father had written at length about the *fluorescence* of a few uranium salts. But Henri occasionally termed the glow *phosphorescence*. In this chapter, to save confusion, the term fluorescence will be used for both phenomena.

In January 1896 Henri Becquerel was one of many hundreds who visited an exhibition in Paris of the first X-ray photographs. He was particularly interested because another scientist had stated that X-rays were produced by the fluorescent glass of the Crookes's tube.[1]

Becquerel then had the idea that if fluorescent glass could produce X-rays so also might other fluorescent substances.[2] He had in mind, of course, the salts of uranium which were by the way of being a family concern of the Becquerels. So he decided to test his idea by making the following simple experiment using the salt called uranium potassium sulphate which he himself had first prepared many years before 1896 for his father's experiments on fluorescence.

The experiment was based on the fact that a photographic plate wrapped in thick black paper is not affected by sunlight but is by X-rays. To the black paper surrounding such a plate he fixed a crystal of his uranium salt. Near to it he fixed a silver coin on top of which he put a similar crystal. Then he placed the plate so that the sunlight would fall on the crystals and make them fluoresce. He expected that the fluorescent crystals would produce X-rays and that those given off by the first crystal would produce a clear impression of the crystal on the developed plate. He also expected that the X-rays given off by the second fluorescing crystal would be stopped by the silver coin so that a dark coin-shaped patch would appear on the developed plate.

When Becquerel developed the plate he found that it looked exactly as he had expected; there was an impression of the first crystal and a well-defined dark patch beneath the spot where the coin had been. It therefore seemed to him that the fluorescing uranium salt gave off X-rays.

On 26 February 1896 he repeated the experiment. He put the wrapped plate out in the open air. The day was dull; so he left the crystals exposed to the light until the following day. This day was also very dull – indeed the amount of sunshine on both days put together had only been enough to make the crystals fluoresce very slightly. So he put the plate, with the crystals and coin still attached to its wrapper, into a dark cupboard with the intention of waiting for a brighter day when he would again expose the crystals. The next two days, as it chanced, were equally dull, so he decided to develop the plate. He expected only a faint impression as a result of the short exposure of the crystals to the sunlight, but to his great surprise, the impression of the crystal and the shadow of the coin on the plate were as clear and as well defined as those given before, even though the crystals had previously been exposed to bright sunlight for a long time!

This experiment seemed to prove that even faintly fluorescent crystals of the uranium salt gave out X-rays. Then he had the brilliant idea that crystals of this salt which had not been made fluorescent (by exposure to light) might give out X-rays. This idea called for further experimentation.

He therefore arranged a photographic plate with crystals and coin attached, as before, but this time he did not expose it to light but put it away in a dark cupboard for a few days. On developing the plate he again obtained the clear impression of the crystal and the dark shadow of the coin. This seemed to show that these crystals gave out X-rays although they had not been made fluorescent. Further experiments not only seemed to confirm this conclusion, but also showed that other compounds of uranium, and even the metal itself seemed to give out X-rays, irrespective of whether or not they were fluorescent.

Then came the other surprising discovery. The rays given off by uranium and its compounds were not X-rays at all, despite the resemblance in their effect on photographic plates! For experiments showed, without the slightest doubt, that they were a kind of rays hitherto unknown. These rays were called 'Becquerel rays' in honour of their discoverer.

Becquerel's chance discovery, as Sir Oliver Lodge said, opened up a new chapter of science. In 1897 Madam Curie began experiments to find out whether other substances gave out similar radiations. She examined almost every known substance and found that a few of them did, so she called them radioactive substances. Her most important discovery, however, was that a known weight of pitchblende, an ore containing uranium, gave off radiations that were more powerful than she expected according to the quantity of uranium in the weight of ore. Hence she suspected that the ore must contain a more radioactive substance than uranium. After a long and tedious process she obtained from a ton or more of ore a very tiny bit of a new, unknown element. She called it radium.

Before Becquerel's discovery, scientists had firmly believed that the atom was the smallest particle of matter and that it could not be divided. Becquerel's news that an element was giving off *something* puzzled the scientists and made them wonder what these rays were made of. By 1902 research had given the answer: these rays contained very tiny particles of matter which must have been split off from the atoms of the elements. Hence it was established for the first time in history that there were smaller particles than atoms and that the atoms of radioactive elements underwent division of their own accord (or, as scientists put it, they underwent spontaneous disintegration).

In the loss of particles by the atoms – for example by radium atoms – it was shown that the energy liberated was tremendous, one estimate being that a gram of radium gave out as much energy as a ton of coal which was burning slowly. But – and this is most important – calculations showed that it took two to three thousand years for this energy to be liberated. Nevertheless, despite the long period involved, it was evident that matter was being changed into energy, a change which was contrary to the views which had been held for centuries.

The impact of Becquerel's discovery was referred to by Sir Henry Dale, who said:[3]

At a meeting of the undergraduates' Natural Science Club in Cambridge in 1897 my contemporary, R. H. Strutt (later Lord Rayleigh, the eminent physicist and co-discoverer of the rare gases), gave us an account of Becquerel's discovery. I well remember the sceptical protest of one of us who was later to become world famous in theoretical physics and astronomy. 'Why, Strutt', he said, 'If this story of Becquerel's were true it would violate the law of conservation of energy.' I like to remember the enterprising spirit of Strutt's rejoinder, 'Well, all I can say is so much the worse for the law because I am quite sure that Becquerel is a trustworthy observer'. And of course none of us there had then an inkling of the enormous expansion of knowledge of which such discoveries as these were to provide the points of origin, or of the whole armoury of physical resources which would thus be brought to the service of medicine.

Thus is was that Becquerel's discovery started research work which led to the discovery of radium, with its great benefits to medicine, and the splitting of the atom, which resulted not only in the production of the atomic bomb but in the gift to man of a tremendous source of energy for peaceful purposes.

It is a staggering thought that these important discoveries had their origin in the fact that the sun did not shine brightly on the last few days of February in the year 1896!

It is a more staggering thought, as mentioned by Professor Strutt, that Becquerel's research was carried out as a result of three false assumptions. The first of these was that X-rays were produced by fluorescent glass – and they are not. The second was that because fluorescent glass supposedly gave off X-rays so would other fluorescent substances – and they do not. The third one was that uranium salts gave off X-rays when they were *not* fluorescent – and they do not. Professor Strutt commented: 'It seems a truly extraordinary coincidence that so wonderful a discovery should have resulted from the following-up of a series of false clues. And it may well be doubted whether the history of science affords any parallel to it.'

41. The Greatest Scientific Gamble in History

ON 6 AUGUST 1945, the first atomic bomb ever used against an enemy was dropped on the Japanese city of Hiroshima, and the destruction it caused was enormous and terrifying. After the war was over, Mr Truman, who was then President of the United States of America, disclosed the war-time secret that the invention and production of this bomb was 'the greatest scientific gamble in history'.

This invention followed the rapid advances made when scientists began to study Becquerel's discovery that atoms of uranium, and other radioactive substances, gradually disintegrate. In simple terms, his discovery that atoms of these substance split of their 'own accord' in nature, led scientists to experiment to see if atoms could be split artificially in the laboratory.

The heaviest naturally-occurring element known was uranium, one atom of which weighs as much as 238 atoms of hydrogen, the lightest one of all. But even the uranium atom is so extremely tiny that millions of them occupy a space no larger than a pin's head. Tiny though an atom is, it is, nevertheless, made up of tinier particles, and may be regarded as being composed of two main parts; there is a central or inner part, which is called the nucleus of the atom and is composed of electrically neutral particles; and an outer part containing electrically charged particles called electrons.

In 1932 a most important experiment was successfully made when two Cambridge scientists, Cockcroft and Walton, showed how atoms could be split in the laboratory. The number which were split in their experiment was relatively very small.

41. The Greatest Scientific Gamble in History

Six years later, two German scientists, Hahn and Strassmann, as a result of their study of uranium, foreshadowed a different kind of splitting from that done in Cambridge. Their work indicated the possibility, in the near future, of splitting the nucleus of each of thousands upon thousands of atoms, one after the other, in the most rapid succession. This new splitting process was given the name of nuclear fission, and the whole process was termed 'a chain reaction'.

It was obvious to scientists that a tremendous amount of energy would be released in a successful chain reaction; indeed, by the year 1939, when the war broke out, the production of atomic energy on a large scale was a distinct possibility in the near future. There had been no secrecy about these discoveries for, in the pre-war scientific world, scientists freely communicated details of their work and discoveries to each other; and it is most likely that if there had been no war, atomic scientists would have concentrated their research on the utilisation of atomic energy in industry. But the coming of World War II completely altered the course of research in Britain and caused the statesmen to give close attention to it.

In April 1940 the British Air Ministry set up a committee of scientists to consider the possibility of producing atomic bombs before the war ended. This committee decided that bombs might be made which would be light enough to be carried by an aeroplane and yet capable of doing as much damage as a bomb containing many thousands of tons of tri-nitro-toluene (or T.N.T. for short), had it been possible that such a weight could be put in a bomb.

The Government accepted this decision and in November 1941 entrusted the research and development to a special wartime department which was called, for reasons of security, the Directorate of Tube Alloys.

The knowledge that the production of such a bomb seemed a reasonable possibility to many of Britain's leading atomic scientists made many of our statesmen and others fear that the German scientists might also produce such a devastating weapon. They knew that German scientists had also been busy on atom-splitting. Indeed, as has already been mentioned, even before the war German scientists had made tremendously important discoveries about nuclear fission. Might they not make other brilliant discoveries during the war which would show them the way to make an atomic bomb? There was another reason for fearing that Germany might be the first to produce this weapon. Uranium is found in only a few places, and one of these is in Czechoslovakia, a country which Germany had overrun.

Therefore many scientists who could be ill spared from other kinds of war work were allocated to atomic research and money was poured out almost regardless of what the total cost would be in the end. In addition, hundreds of skilled engineers and other craftsmen were taken from other war work to assist in this vital research.

Another important consideration was that a most important substance for atomic research was produced only in Norway. This substance, which is called heavy water and is related to ordinary water, was then obtained by using specially designed apparatus and at a very slow rate, drop by drop. It was manufactured only at the works of the Norsk Hydro Company in Norway.

Early in the year 1940 the French government entered into negotiations with the management of this company for the purchase of all the heavy water then in store. The managers agreed to sell on condition that the greatest secrecy was observed for fear of German reprisals at some later date. Thus practically all the heavy water in the world was taken to France. It arrived there just in time, for Norway was overrun by the Germans only a few weeks after its sale.

But the heavy water soon had to make another journey, for in June 1940 France, in its turn, was also overrun by the Germans. Fortunately a few leading French scientists managed to escape. They made their way in secret to a French port, taking with them all the precious heavy water, about 165 litres, or twelve gallons. A British ship was docked there and, as soon as the men and their cargo had got on board it, it set sail and reached Britain safely. The heavy water was then taken to the Cavendish Laboratory in Cambridge where it was of very great value in research projects.

When the Germans overran Norway their scientists had available, of course, the heavy water which the Norsk Works continued to produce. So, in the winter of 1942-3, the Allied leaders decided to hinder its production. A bold plan of attack was drawn up. It was based on the information given by Norwegian refugees who had worked at Norsk and who knew the location in the works of the vulnerable and important parts of the apparatus. This information was studied by a small party of Allied and Norwegian commandos, trained in sabotage. A first raiding party was sent from Britain, but the raid was not successful. So a second raid was then made.

This was led by a young Norwegian officer, Lieut Haugen. The nine Norwegians composing it were dropped from a Stirling bomber by parachute near the works. They made their way over a frozen river, evaded the German guards at the Norsk Hydro works and forced an entry. Down in the basement they put high explosives beneath the most important parts of the apparatus and lit the fuse. The explosion not only wrecked most of the apparatus, it destroyed the store of six months' supply of heavy water. Indeed the works were sabotaged so successfully that the Germans had not restored them by the time of their capitulation.

Many months later, when this capitulation was drawing near, the Allies decided to preserve as much of the works as had been repaired. The young officer Haugen again landed in Norway by parachute. There he organised a thousand men of the Home Army with weapons dropped by British planes. Two days before the Germans were due to leave he stormed the works. The guard surrendered, although they had been given orders to blow up the works.

Another daring venture was made in November 1943, when the Danish underground movement learned that the Nazis had decided to round up Danish Jews and had ordered the arrest of Neils Bohr, who was professor of physics at Copenhagen University and one of the world's foremost workers on atomic energy. The underground leaders arranged for him to escape from the

Gestapo by fishing boat. He landed in Sweden where the local police, who had previously agreed to guard him from German agents, watched over him at every place through which he passed. Finally he boarded a British plane and reached Britain safely. There he immediately began to aid the Allies in their atomic research.

Allied intelligence was by then sending frequent news of places in Germany and occupied countries where atomic research was probably being carried out and arrangements were made for these places to be attacked by Allied bombers.

The British expected that the Germans in their turn would attack atomic research posts in Britain, so in 1942 the research was transferred to America, where work proceeded at a very rapid pace. A wonderful organisation was set up, with numerous teams of scientists at work in many laboratories, all busy on their own special bit of research; hundreds of work people were making different things or working on some job or another without knowing what the others were doing. All this was carried on under the direction of a very small committee whose members alone knew all the secrets and who had constantly before them the fear that the Germans might beat them in the race to produce the atomic bomb.

The problem confronting the Allied scientists had not been solved by D-Day (the first day of the invasion of Normandy by the Western Allies). There was then much speculation amongst the Allied leaders about the progress the German scientists might have made. Possibly some clues might be obtained from the occupied territories. To investigate this possibility a team of scientists was landed in France on the day after D-Day. They were ordered to advance just behind the front-line troops and to seek evidence of German progress in atomic research. One of the methods they used is illustrated by what they did on reaching the River Rhine. They knew that the Allied scientists in America used a large tower called a uranium pile which had to be kept cool by cold water and was built therefore on the banks of large rivers. During its passage through the pile the water became radioactive.

41. *The Greatest Scientific Gamble in History*

The Allies assumed that if the Germans had made much progress in their atomic research they might also be using uranium piles. Hence the scientists who advanced behind the front-line troops were instructed to take samples of water from all the big rivers in former occupied territory and in Germany itself and test it for radioactivity. All the tests showed the absence of radioactivity. It seemed that the Allies were much further advanced in the race to use atomic energy or, as Mr Churchill has said, 'By God's mercy British and American science has outpaced all German effort'.

Gradually it then became known that the Germans had not made as much progress towards making an atomic bomb as the Allies had feared, for most of their research on atomic energy had been concentrated on its application to industry.

There were several reasons for this lack of progress and the attitude of Dr Otto Hahn is worthy of note. As has already been mentioned, he was one of Germany's prominent atomic scientists in 1939 and there is little doubt also that he was one of the few Germans who could have led a team of atomic research workers in a successful effort to produce a bomb. During the war he did, indeed, carry out much important work on atomic energy at the Kaiser Wilhelm Institute for Chemistry in Berlin and by the year 1942 he had found out that it was practical to obtain power by utilising atomic energy and had even thought of an atomic bomb.

His book, written in 1950, mentions that he realised that 'his country was ruled by Hitler in whose hands atomic energy might well have spelt the doom of all mankind'. Hence 'during the whole of World War II he directed his work and that of his collaborators towards its utilisation for peaceful purposes only'.

Lack of equipment, he admitted, forced his hand to some extent and, as things turned out, it would have been very difficult for Germany, especially in the last months of the war, to do a great deal because of their lack of technical means and the frequent bombardment which their factories sustained from the air.

The Allies continued their efforts to produce a bomb after the capitulation of Germany and the first one was ready for its tests in July 1945. It was a terribly destructive weapon with more power than 20,000 tons of T.N.T. and more than two thousand times the blast-power of the largest bomb ever used before. These first bombs had cost five hundred million pounds to produce, and as many as 125,000 people had been employed at one time or another on the project, many of them for as long as two and a half years.

The war with Germany had ended in May but Japan was still fighting the Allies and since the bomb was ready someone in authority had to decide whether to use it. Mr Churchill and President Truman met at the Potsdam Conference and decided to do so; they informed Marshal Stalin, the leader of the Soviet Union, that 'an explosive of incomparable power' was going to be used against Japan.

It was agreed first to give the Japanese Government an ultimatum to surrender but also to warn them that the alternative to unconditional surrender was complete and utter destruction of their cities. This the Allies did, but the Japanese Prime Minister refused these terms, while privately continuing his efforts to negotiate through Marshal Stalin.

The test on 16 July had been successful but the Allies then possessed only two atomic bombs and it would have taken a long time to make others. So these two were the only ones which could be used for many months. In the short space of a few days the bombs were hurried across the Pacific and all was ready for them to be dropped on 6 August 1945.

That day began as a sunny morning in the Japanese city of Hiroshima, a city which was a fortified port on the same island as Tokyo; it was one of the chief supply depots of the army and had ship-building yards, cotton mills and war factories. The attack came without warning and caught everyone by surprise. Within a minute many thousands of men, women and children, were horribly killed, most of them being burnt to death by the tremendous heat of the explosion; and the heart of the city was wiped out as though some great bulldozer had swept over it.

The same day, Mr Truman, in a broadcast speech, told the Japanese that if they did not accept the peace terms of the Allies, they could expect a rain of ruin from the air the like of which had

41. *The Greatest Scientific Gamble in History*

The atomic bomb is dropped

never been seen on this earth. Mr Churchill also broadcast in similar terms.

But no offer of surrender was received and so the second bomb was dropped, only three days later, with similar awful results, on the city of Nagasaki.

The second blow broke the nerve of the Japanese Government. The slaughter at Nagasaki had been as appalling as at Hiroshima. No precise figures were published, but an estimate given by Tokyo radio said there had been 280,000 casualties (of these about 105,000 had been killed and the others injured) in the two cities. The Emperor was able to get a majority support in the War Cabinet for the unconditional surrender he had long privately pressed for. On 10 August a broadcast from Tokyo announced that the Japanese Government 'earnestly desires to bring about an early termination of hostilities with a view to saving mankind from the calamities to be imposed on them by the further continuance of the war'.

It is now known that the fear which spurred on the Allies in their endeavours to outpace the German scientists was greater than it need have been had they known what was actually happening in atomic research in Germany. Nevertheless, the Allied leaders most certainly were not only wise to fear the foe but had many reasons to believe that the Germans could and would produce an atomic bomb.

The scientific gamble, however, was not a profitable one so far as the waging of the war against Germany was concerned. The diversion of scientists from such work as radar, the detection of magnetic mines and submarines, the preparations of a hundred and one things for the invasions, was done at great cost to the war effort. Yet Germany was defeated before the bomb was produced.

It is a fact that the Japanese leaders capitulated only four days after the first atomic bomb had been dropped. But it is widely agreed by experts that Japan would have surrendered in any case before the autumn. Whether the atomic bomb should have been used or not will be disputed for many years. But these words of Mr Churchill, written on the fateful 6 August 1945, will be echoed by many people: 'This revelation of the secrets of nature, long mercifully withheld from man, should arouse the most solemn reflections in the mind and conscience of every human being capable of comprehension. We must indeed pray that these awful agencies will be made to conduce to peace among the nations, and that instead of wreaking measureless havoc upon the entire globe, they may become a perennial fountain of world prosperity.'

42. Some Early Steam Engines

LITTLE PROGRESS was made in the invention of an engine which would work by steam until the seventeenth century, when one of the outstanding contributors was Edward Somerset, Second Marquis of Worcester.

The marquis had fought with Charles I's army, for which act he had been banished, by resolution of Parliament, and condemned 'to die without mercy if ever he were found within the

42. *Some Early Steam Engines*

limits of this nation'. Despite this resolution he returned as a royal spy. He was arrested in the year 1652, and as was so often the case in those days, he was committed to the Tower of London without a trial and left there.

Before becoming a soldier the marquis had been most interested in the science of his day; hence, during his stay in prison, which lasted for two years, he considered various scientific problems. One day, according to a legendary story, whilst he was cooking his dinner he observed that the lid of the pot was continually being forced upwards by the vapour of the boiling water. The story continues:

Being a man of thoughtful disposition, and having a taste for scientific investigation, he began to reflect on the circumstances. It occurred to him that the same power which was capable of raising the iron cover of the pot might be applied to a variety of useful purposes.

When he obtained his freedom he used this idea to design a steam engine to pump water out of mines.[1]

There is, however, no conclusive evidence to prove that he ever made a steam engine; but he did outline how to make such a machine in his celebrated book entitled *A Century of Inventions*.

*   *   *   *   *

The second story of this series concerns a military engineer named Thomas Savery who spent most of his leisure time carrying out mechanical experiments. Savery was interested in methods of pumping water from the mines in Cornwall and invented a steam engine for this purpose.

He has been accused of making his steam engine to the plans described in the Marquis of Worcester's book, and his accuser gave this account of how Savery tried to conceal what he had done:

He bought all the Marquis of Worcester's books that he could purchase and burnt them, the better to conceal that he had copied from them. Then he announced that he had found out the power of steam by chance and invented the following story to persuade people to believe him. One day having drunk a flask of wine in a tavern he threw the flask on the fire. [A little wine was left in the bottom of the flask and the heat changed it into vapour which forced the air out of the flask.] Then Savery, happening to glance at the fire, saw that the flask was full of steam. He snatched it off the fire and plunged it mouth downwards into a basin of cold water. Immediately some of the water in the bowl was forced into the flask.[2]

Savery's fire engine, as it was called, resembled the flask and bowl of water in principle. It consisted mainly of a large globe connected with a long pipe, which reached the water at the bottom of the mine. This globe was first filled with steam, which was then condensed to a few drops of liquid by the application of cold water to its outer surface, hence a partial vacuum was formed inside. Immediately water rushed up the pipe and into the empty space. The water was then poured out. The processes of filling the globe with steam, condensing the steam, and emptying the globe of water, were repeated as often as necessary.

Whether Savery's discovery took place as a result of the flask incident or not, it must be admitted that the description given by the Marquis of the method of making a steam engine was far too scanty to enable anyone to make an engine from it. In any event, Savery's steam engine was soon displaced by a superior one designed by Thomas Newcomen, a blacksmith of Dartmouth.

A legend in Dartmouth closely resembles the one about the Marquis. It relates that Newcomen was sitting near the fire one day when he noticed that the escaping steam repeatedly lifted the lid of the kettle. This convinced him that escaping steam could exert power and led him to design his famous engine.

The long beam of this engine was balanced on a pivot (A), and a weight (B) was hung from one end of it so that the beam moved easily up and down, like a see-saw. A chain was fastened to the other end as well as to the piston (P). Hence when the beam moved up it lifted the piston to the top of the cylinder. A boy was in constant attendance to open and close the taps: as soon as the piston had reached the top of the cylinder he opened tap (C), thereby admitting steam until the cylinder was full of it. Then he closed the tap and opened tap (D), thus admitting a jet of cold water. This caused the steam to condense, producing a partial vacuum in the cylinder. Hence the pressure of the air on the

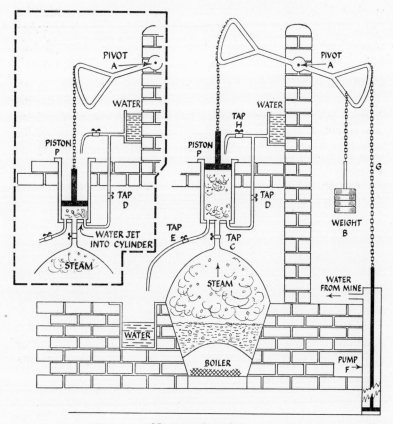

Newcomen's engine

outer surface of the piston was sufficient to force the piston to the bottom of the cylinder. The cold water was run out of the cylinder by tap (E) and all was then ready for the process to be repeated.

As will be obvious from the diagram, the beam in its up-and-down movement worked the pump (F) by means of the chain (G).

The first machine made by Newcomen was a little different from this description, for the steam was condensed not by a jet of water at (D), but by cold water run from the tap (H) into the top of the cylinder above the piston. One day, to his great surprise, the engine suddenly began to go very fast, giving several

123

strokes in the time formerly taken by one. On investigation he found that a hole had been worn in the piston and cold water was dripping into the bottom part of the cylinder through this hole and thus condensing the steam. 'When this happened', according to one of his biographers:

A new light suddenly broke upon Newcomen – the idea of condensing the steam by injecting cold water directly into the cylinder instead of applying it to the outside at once occurred to him; and he proceeded to embody the expedient, which had thus been accidentally suggested, as part of the machine. He installed the tube (D) and fitted it with a rose head, so that a shower of cold water could be injected into the cylinder before each descent of the piston. The steam was condensed almost at once and the down stroke thereby made almost immediately following the upstroke.

This new engine was in use from about 1711 for many years. It was tended by a boy who had to open and close the taps at the proper times. It is said that one such boy, named Humphry Potter,[3] either soon got tired of this monotonous job, or else wished to play marbles upon the engine-room floor – a place which was well suited to that game. Other boys came to jeer at him, and played marbles whilst he worked until he could stand it no longer. Then it happened that one day when his master entered the engine room he saw young Humphry playing marbles. We are told that:

His first duty, that of punishing Humphry, was strenuously performed, and only then did he observe that the pump engine, even though it was unattended was faithfully performing its duty. Soon he saw that the ingenious boy had fastened bits of sticks and strings of the proper length to the taps and the beam so that the taps were opened and closed at the correct time by the rise and fall of the beam.

His master, Henry Beighton, immediately realised the brilliance of the boy's idea and before long had converted the strings and catches of the boy into metal rods and tappets. Hence the steam engine, through a boy's ingenuity, aroused by his desire to play marbles, became self-working, or automatic, the only man in attendance on it being the 'firer-up' or stoker.

The person who first told this story was well acquainted with Henry Beighton from whom he got his information; but there appears to be no confirmatory evidence.

42. Some Early Steam Engines

Newcomen's engine was widely used, but it was somewhat slow in action; it was also wasteful of fuel mainly because of the way used to cool the cylinders and so to condense the steam. In 1765 James Watt, then a young man engaged in instrument-making at Glasgow University, was given the job of repairing a model of Newcomen's engine. He studied it very carefully and noticed the defects.

For a long time he turned over in his mind methods of improving it, but could think of no way of doing so until one Sunday, early in 1765, when, to use his own words, his brilliant idea was born as follows:

I had gone for a walk on a fine Sabbath afternoon. I had entered the Green by the gate and had passed the old washing-house. I was thinking of the engine at the time and had gone as far as the herd's house when the idea came into my mind. I had not walked further than the Golf House when the whole thing was arranged in my mind.[4]

He got up early next morning and tried out his new plan. It was a simple improvement, consisting of another vessel connected to the cylinder, in which the steam could be condensed. Thus the cylinder itself did not need to be cooled. The use of the condenser improved the efficiency of the engine and resulted in a great saving of fuel.

The traditional story of Watt and the kettle is well known. It was first told about half a century after the event was supposed to have occurred. According to this story, young James was sitting one evening with his aunt, Miss Muirheid, at the tea table, when she said:

James Watt, I never saw such an idle boy; take a book or employ yourself usefully, for the last hour you have not spoken one word, but taken off the lid of that kettle and put it on again, holding now a cup, now a silver spoon over the steam, watching how it rises from the spout, and catching and counting the drops of hot water it falls into. Are you not ashamed of spending your time in this way?[5]

Another version of this story relates that he closed the end of the spout to stop the steam escaping and noticed that the steam then lifted the lid off the kettle. It is noteworthy that stories of the kettle lid and the power of steam are associated with Watt, Newcomen and the Marquis of Worcester.

Watt's new engine was a tremendous advance on former designs. He entered into partnership with an industrialist named Boulton, and soon they had set up a large works for manufacturing steam engines. They became famous and Boulton was invited to a court levée. There George III graciously questioned him about what he did, 'I am engaged, your Majesty, in the production of a commodity which is the desire of Kings'. 'And what is that?', asked the King. 'Power, your Majesty', replied Boulton.

Boulton was, of course, referring to the kind of power required to do the sort of work done by horses before the invention of the steam engine. It seemed to his partner that an obvious way of comparing the performance of the newly invented engines was to estimate the number of horses that would be required to do the same work as an engine. This Watt did in his usual efficient manner. First he obtained permission from a firm of brewers to experiment with some of their heavy dray horses at a brewery in London. A weight of one hundred pounds, attached to a long rope, was put at the bottom of a deep well, with the free end of the rope passing over a pulley at the top.[6] This free end was attached to a horse. Watt found that, on the average, a horse could walk on the level ground at a speed of two and a half miles an hour whilst pulling the weight to the top of the well. That is, the horse was travelling at the rate of two and a half miles an hour, which is 220 feet a minute, whilst raising a weight of a hundred pounds.

Thus, the horse raised a weight of 100 lbs through a height of 220 feet in a minute. Mathematically, this is equivalent to raising 1 pound through a height of 22,000 feet. The experiment showed that horses like the ones Watt was using could each do 22,000 'foot-pounds' of work per minute. He knew that the horse was not pulling directly on the weight, and that the friction of the pulley would have some retarding effect. He also knew that the horses used in the experiment might have been less strong than other horses, and so on. To be on the safe side, therefore, he decided to increase his result of 22,000 by fifty per cent, bringing it to 33,000. Therefore, as physicists now say, 1 horse power = 33,000 ft lbs/min. Thus, when Watt gave the purchaser details of the horse-power of an engine the purchaser knew that for each 'one horse-power' of the engine Watt guaranteed that it would raise 33,000 pounds to a height of one foot in one minute. The purchaser, therefore, had a way of comparing the performance of the engine with the work done by horses.

43. On the Road

BEFORE THE end of the nineteenth century inventors like Savery, Newcomen and Watt had produced steam engines which could replace the horse in such work as pumping water from the mines. Other inventors had turned their thoughts towards steam engines or 'locomotives' which would also replace the horse in pulling carriages and coaches along the roads.

One of the first inventors of a steam locomotive was a Frenchman named Joseph Cugnot. He was a military engineer who succeeded first, in 1769, in making a small working model of a steam locomotive, and later in making a full-scale engine which would move along the road without rails. This full-sized engine had three wheels, one in front and two behind, and was fitted with a two-cylinder engine which acted on the front wheel. The boiler produced only enough steam to run the locomotive for a quarter of an hour, after which the machine had to stop until more steam had been generated.

A trial was ordered by the Minister of War, who was greatly interested in the possibility of using locomotives instead of horses for pulling the guns.

On the day of trial the engine was taken to one of the streets of Paris, along which it travelled at ten miles an hour, and all was well until it came to a corner. Then, as it turned it toppled over with a loud crash. The engine was damaged and some accounts say that a few people were hurt, including some of the very important military men who had assembled to watch the trial.

The trials were stopped immediately and the engine locked up so that Cugnot could not use such a dangerous thing again (it is said that Cugnot himself shared imprisonment with his engine). However, the engine was not destroyed, and one writer states that later Napoleon became greatly interested in it, as he was in anything likely to be put to good use in his army. In 1801 he decided to give it another trial, but before this could take place he had set sail for the invasion of the Nile. The 'resting place' for this first engine is a suitable one, for it is placed in a museum near the spot where it was first put on trial.

*   *   *   *   *

A more successful attempt to invent a road locomotive was made by the young Scot, William Murdoch, who was then residential engineer in Cornwall for the important firm of Boulton & Watt, manufacturers of stationary steam engines for work in mines (see chapter 42). Murdoch, who had a very good knowledge of the steam engines which his firm produced, decided to make a small model of a locomotive which would move along a road under its own power. He completed this in 1784.

It was about nineteen inches long and fourteen inches high, with one wheel in front and two behind; the copper boiler was heated by a spirit lamp. The cylinder was three-quarters of an inch in diameter with a two-inch stroke. The engine worked by the expansive force of steam, which was discharged into the atmosphere after it had done its work in the cylinder.

First of all he tested his model indoors and found that it could draw a small model waggon around the floor of his lodgings in Redruth. Then he decided to try it on the fairly smooth surface of the church drive. The following story of this 'run' is told by F. Trevithick:[2]

Murdock, in the dark and in secret made a model steam locomotive designed to travel on the roads; it was too small for a driver to be carried by it. Notwithstanding its diminutive dimensions, this little gentleman managed to outrun the inventor on one occasion. One night, after returning from his duties at the mine in Redruth, Cornwall, he wished to put to the test the power of his engine, and as railroads were then unknown, he had

43. On the Road

recourse to the walk leading to the church, situate about a mile from the town. This was very narrow, but kept rolled like a garden-walk. The night was dark, and he alone sallied out with his engine, lighted the fire (a lamp under the boiler), and off started the locomotive, with the inventor in full chase after it. Shortly after he heard distant despair-like shoutings; it was too dark to perceive objects, but he soon found that the cries for assistance came from the worthy pastor who was at that moment going into the town on business.

For the vicar had suddenly seen something moving towards him in the darkness with great speed, spurting flames of fire and hissing and spitting at him. 'He imagined it was the Evil One himself and shouted loudly for help.'

Murdoch chases his engine

Murdoch appeared quickly on the scene and, seeing the vicar scared out of his wits, began to soothe him by explaining what the strange moving object was. He then ran after his model and retrieved it before it could scare any one else.

News of Murdoch's experiments reached James Watt and he became concerned lest Murdoch should spend too much time on his model locomotives to the detriment of his work.[3] Accordingly he asked Boulton to persuade him to give up his experimental work. He was just in time, for Murdoch was on his way to London to take out a patent for making locomotives when Boulton met him and managed to persuade him to return to Cornwall.

On reaching home Murdoch unpacked his model engine and set it going, with regretful thoughts that this was almost the last time he would run his engine; but before he put it away he showed Boulton that it could pull a load containing the fire-shovel, the poker and tongs.

One of the models – for Murdoch made two or three different ones – was preserved by the family for a hundred years and then it was purchased for a museum.

*   *   *   *   *

About this time, in a village not far from Murdoch's, there lived another engineer, Captain Richard Trevithick, who was also employed in the mines. He too was of an inventive turn of mind and designed an engine in which steam at high pressure was used to force the piston to the cylinder head. The steam was then allowed to escape and made a 'puffing' noise. This noise caused the local people to call the engine a 'puffer' and the nick-name stuck to all steam locomotives for at least a century.

His first road locomotive was assembled in the local smithy: it had a cylinder, horizontal boiler and chimney, and was large enough to carry at least six men. All was ready for a trial run on Christmas Eve, 1801.

Old Stephen Williams, in 1858, recollected the events of that day:

I knew Captain Dick Trevithick very well; he and I were born in the same year. I was a cooper by trade and when Captain Dick was making his first steam-carriage I used to go every day into John Tyack's blacksmith's shop at the Weith close by here, where they put her together. There was a deal of trouble in getting all the things to fit together.

In the year 1801, upon Christmas Eve, coming on in the evening, Captain Dick got up steam, out in the high road, just outside the shop at the Weith. When we see'd that Captain Dick was agoing to turn on steam, we jumped up as many as could; maybe seven or eight of us. 'Twas a stiffish hill going from the Weith up to Camborne Beacon, but she went up like a little bird.

130

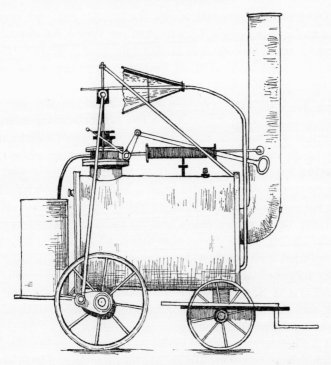

Trevithick's engine

When she had gone about a quarter of a mile, there was a roughish piece of road covered with loose stones; she didn't go quite so fast, and as it was raining, and we were very squeezed together, I jumped off. She was going much faster than I could walk, and went on up the hill about a quarter or half a mile further, when they turned her and came back again to the shop.[4]

Another of the 'hitch-hikers' recalled that

Trevithick called upon the people to 'jump up' so as to create a load on the engine and it soon became covered with men, who did not seem to make any difference to the speed, so long as the steam was kept up. The double-acting bellows, turned by the engine, fanned the fire, and the engine was called Captain Dick's puffer, from the steam and smoke which puffed out of the chimney at each stroke of the engine.

Trevithick used a difficult road for this first journey. It had a gradient of about one in fifteen and the surface was rough and stony. Horse-drawn vehicles travelling up the hill could go only at a slow walking pace.

The next day Trevithick took it for a 'short mile', passing the house of a friend of his, Captain Andrew Vivian. An old lady of the household was amazed at the sight, and cried out, 'Good gracious, Mr Vivian, what will be done next? I can't compare un to anything but a walking puffing devil'.

On 28 December 1801 the engine began its last journey; Trevithick, Captain Vivian and a few others were on board. It travelled well until it came to a 'kind of open water-course across the road', but in crossing this it bounced about; the steering handle was jerked out of the driver's hand and the engine overturned. This was after it had travelled very well up the hill for three or four hundred yards.

There was a public house nearby, and the carriage was put under shelter whilst the parties adjourned inside. There they 'comforted their hearts with a roast goose, and proper drinks'. They forgot all about the engine and that they had left the fire burning in it. Soon the water boiled away, 'the iron became red hot, and nothing that was combustible remained either of the engine or the house'. One legend has it, however, that the engine was deliberately set on fire, whilst its driver was in the hotel, by coach-drivers who were afraid that they and their horses would lose their jobs if the engine proved successful.

Trevithick and his colleagues set to work and made another engine and in 1802 obtained a patent for 'moving-carriages by steam'. They were encouraged to take an engine to London for exhibition, one account stating that 'en route it was driven by its own engine to Plymouth, ninety miles from Cambourne, and then shipped by water'. In London a new passenger carriage was made but the engine ran for a short time only and then back to Cornwall it went. The experiment of using a steam locomotive to convey passengers along the public highway had not proved as successful as the inventor had hoped.

Trevithick was thoroughly disappointed and emigrated to South America, where he worked at his former job as a mining engineer. Later he returned to England, where 'he died on the Parish', so poor was he. His former workmates, however, subscribed enough money to give him decent burial.

In the year 1932 a statue was erected to his memory; it faces Beacon Hill, which was the first hill up which his steam locomotive travelled.

44. Two Young Men Find Employment

IN THE year 1791 a Yorkshire blacksmith removed to London and there fell ill and died shortly before the birth of his son, Michael Faraday (chapters 26 and 44). Hence his family was left almost in poverty and Michael had to earn his living at an early age.

When he was thirteen he was working as an errand boy for a bookseller and his main job was to deliver newspapers. (In those days a few papers were loaned to customers. Michael delivered the paper at a house, left it there for an hour until his customer had read it and then called to collect and deliver it to another customer, and so on.) A year later he became apprenticed to a bookbinder, a job which had an important bearing on his future.

With the eagerness of a keen student young Michael read many of the books which came his way while he was working in the book shop, but one in particular made a big impression on him. This was Mrs Marcel's *Conversations on Chemistry*, a book then widely used in the teaching of chemistry; and it gave the boy his first taste of science. Before long he began to attend evening lectures on this subject and not only made careful notes of the lectures but bound his note-books into volumes.

Then came a red-letter day in his life, when one of his employer's customers took him to hear four public lectures on science by the famous Sir Humphry Davy. Sir Humphry was then drawing fashionable audiences to the Royal Institution to hear his eloquent and interesting lectures in chemistry. Faraday followed his usual custom of taking full notes of these lectures and afterwards wrote them out neatly. He then bound the sheets together, the whole making a beautifully bound book of 386 small, quarto pages.

The next step in his career has been described by Faraday himself:

My desire to escape from trade, which I thought vicious and selfish, and to enter into the service of science, which I imagined made its pursuers amiable and liberal, induced me at last to take the bold and simple step of writing to Sir H. Davy expressing my wishes and a hope that, if an opportunity came his way, he would favour my views: at the same time I sent the notes I had taken of his lectures.[1]

Davy received this letter just before the Christmas of 1812 and showed it to a friend who happened to be visiting him at that time, commenting, 'What am I to do, here is a letter from a young man named Faraday: he has been attending my lectures, and wants me to give him employment in the Royal Institution – what can I do?' 'Do', replied his friend, 'put him to wash bottles: if he is good for anything he will do it directly, if he refuses he is good for nothing'. 'No, no', replied Davy, 'we must try him with something better.'[2]

Davy's interest in Faraday had been whetted and he wrote a kind reply to the youth of twenty-one, promising him an interview at the end of January. Accordingly, Sir Humphry saw Faraday but had to tell him that there was no vacancy and advised Faraday to stick to his present job. For science, he said, 'was a harsh mistress', and, from a financial standpoint, 'poorly rewards those who devote themselves to her services'. Nevertheless, he promised to send Faraday all his bookbinding work.[3]

Faraday was greatly disappointed when his efforts to become a 'full-time' scientist seemed to have come to nought. Then came his opportunity, through a cause which no one could have predicted. It happened that at the beginning of the year 1813 the laboratory boy at the Royal Institution, William Payne, was promoted to help the instrument maker, Mr Newman, his chief job being to clean and repair apparatus. Payne and Newman did not get on well together, and one night the Superintendent heard a great noise in the lecture room. He hastened to the room to find the two bitterly quarrelling. Newman had accused the

44. *Two Young Men Find Employment*

younger man of neglecting his duty and Payne had struck him. The Superintendent stopped the quarrel and later reported it to the managers. As a result Payne was dismissed.

Sir Humphry then recalled the application which young Faraday had made for a job and decided to offer him the vacant post. There was a touch of the dramatic in the way Faraday received the news of his appointment, the account of which runs:

One night when undressing in Weymouth Street, Faraday was startled by a loud knock at the door; and on looking out he saw a carriage from which the footman had alighted and left a note for him. This was a request from Sir H. Davy that he would call upon him the next morning. Sir H. Davy then referred to their former interview and inquired whether he was still in the same mind, telling him if so he would give him the place of assistant in the laboratory of the Royal Institution: the salary was to be 25/- a week with two rooms at the top of the house.[4]

Faraday gladly accepted the post. His duties were:

To attend and assist the lecturers and professors in preparing for and during lectures. When any instruments or apparatus may be required, to attend to their careful removal from the model room and laboratory to the lecture room, and to clean and replace them after being used, reporting to the managers such accidents as shall require repair, a constant diary being kept by him for that purpose. That on one day in each week he be employed in keeping clean the models in the repository, and that all the instruments in the glass cases be cleaned and dusted at least once within a month.[5]

Faraday did not stay long in such a humble but necessary job. His ability was so outstanding that Davy and others quickly found that he was fit for better work. He rose higher and higher until twelve years later he had succeeded Sir Humphry, becoming the Director of the laboratory at the Royal Institution. There, for the next forty years, he carried out those brilliant investigations in science which have conferred lasting benefits on the human race.

*   *   *   *   *

The appointment of the young engineer named William Murdoch (chapter 43) was as romantic as that of Michael Faraday. William was born in the year 1754 in a small village in Ayrshire.

His father was a farmer and wheelwright. William, who inherited his father's craftsmanship, followed the trade of wheelwright until he reached the age of twenty-three.

As a boy he had done some rough and ready experiments on heating a crude kind of coal (chapter 13) and this early inventiveness coupled with skill with his hands led him to choose engineering as his life's occupation.

There was little opportunity for a young and ambitious engineer at home, the Mecca of such being the large works of Boulton and Watt in Birmingham.

William decided to go to Birmingham to see if this progressive firm would employ him and told a friend of his intention. His friend advised him to wear a bowler hat 'because all prosperous young men in the south wore such a hat'. So William acquired one, and we must picture him wearing a bowler as he walked most of the way from Scotland to Birmingham, the fare of a stage-coach the whole of the way being more than he could afford.

On his arrival at the works he asked to see Mr James Watt; but Watt was away from home that day so the youth was shown into Mr Boulton's office. Boulton at first did not reply at all encouragingly to young Murdoch's application for a job, for it chanced that at that particular time business was slack and there were no vacancies. However, Mr Boulton was a humane man and on learning that William had travelled such a long way in search of work began to have a conversation with him. William, like most country boys, was somewhat shy on being addressed by such an important person and had some difficulty in knowing what to do with his hands. In a few moments he had begun, unconsciously, to twist his bowler hat in them.

The employer noticed the hat and saw that it was not made of felt or of cloth, which were the usual materials. It looked as if it had been made of something else and had been painted. Boulton's and Watt's biographer, Smiles, described what then happened:

'That seems a curious sort of hat' said Boulton, looking at it more closely; 'why, what is it made of?' 'Timmer, sir', said Murdoch modestly. 'Timmer,

Murdoch and his wooden hat

do you mean to say that it is made of wood?' 'Yes, sir'. 'Pray, how was it made?' 'I turned it mysel', sir, in a bit lathey of my own making'.

Boulton looked at the young man again. He had risen a hundred degrees in his estimation. He was tall, good-looking, and of open and ingenuous countenance; and that he had been able to turn a wooden hat for himself in a lathe of his own making showed that he was a mechanic of no mean skill.[6]

This was proof enough for Boulton, that rare judge of men. Here was a natural-born mechanic, certain. The young man was promptly engaged for two years at fifteen shillings a week. His history is the usual march upward until he became his employers' most trusted manager in all their mechanical operations.[7]

45. Courtesies to Enemy Scientists

IN THE two world wars of this century scientists of the opposing nations put their knowledge at the disposal of their own countries and worked in conditions of the greatest secrecy on problems such as those mentioned in this book: the invention and manufacture of the first tanks (chapter 92); the production of the first atomic bomb (chapter 41) and the use of poison gas (chapter 21). Secrecy was, of course, secured and maintained in hundreds of other projects and there was no communication between enemy scientists in either war. Indeed any attempt at such communication would have been reckoned as treason of the highest degree.

This attitude towards fellow scientists in enemy countries has not, however, always been taken, as an interesting communication written in the year 1803 shows. In that year England and France were bitterly opposed in the Napoleonic Wars and feelings ran high. The President of the Royal Society in England was Sir Joseph Banks and he wrote as follows to his French equivalent, the President of the Institut National:[1]

If I cannot maintain a correspondence with learned Englishmen in France, without being accused of employing them for political objects and if gentlemen of known reputation and honour cannot visit your country for the purpose of giving and receiving scientific information without being exposed on every turn of public affairs to the vile imputation of acting as spies, – it will be impossible for me to preserve a constant intercourse of good office between men of science in the two nations.

But it must be added that Sir Joseph's dealings with France during the war by no means met with universal favour in England.

*　　*　　*　　*　　*

In 1779, a few years after the English colonists in North America had proclaimed their independence (page 79), Sir Joseph complimented another enemy of his country.

For some years the American colonies and the mother country had been fighting each other, so that by 1779 the two opposing peoples were bitterly hostile. American small ships (privateers) were deliberately based on France, which country was then also at war with England, to enable them to raid and harass British shipping.

45. Courtesies to Enemy Scientists

Between the years 1768 and 1779 Captain Cook was sailing the southern seas, exploring new lands in what is now called Australasia. He had been at one time an acquaintance of Benjamin Franklin, who was not only a scientist but also a leading statesman of America, and at that time the representative of the United States at the Court of France (chapter 34). Franklin's letter, which follows, describes the action he took on the 10th day of March 1779.

To all captains and commanders of armed ships, acting by commission from the Congress of the United States of America, now at war with Great Britain:

GENTLEMEN: A SHIP having been fitted out from England before the commencement of this war, to make discoveries of new countries in unknown seas, under the conduct of that most celebrated navigator and discoverer, Captain Cook – an undertaking truly laudable in itself, as the increase of geographical knowledge facilitates the communication between distant nations in the exchange of useful products and manufactures, and the extention of the arts, whereby the common enjoyments of human life are multiplied and augmented, and science of other kinds increased, to the benefit of mankind in general. This is therefore most earnestly to recommend to every one of you, that in case the said ship, which is now expected to be soon in the European seas on her return, should happen to fall into your hands, you would not consider her as an enemy, nor suffer any plunder to be made of the effects contained in her, nor obstruct her immediate return to England, by retaining her or sending her into any other part of Europe or to America, but that you would treat the said Captain Cook and his people with all civility and kindness, affording them, as common friends to mankind, all the assistance in your power which they may happen to stand in need of. In so doing you will not only gratify the generosity of your own dispositions but there is no doubt of your obtaining the approbation of the Congress and your other American owners.

I have the honour to be, gentlemen,

Your most obedient, humble servant,

B. Franklin.

At Passy, near Paris, this 10th day of March, 1779.[2]

Franklin issued this letter on his own authority but later Congress (the United States 'Parliament') agreed to support it.

Captain Cook, however, had been killed by the natives of one of the newly discovered islands before Franklin's gift of a passport had reached him.

After the news of Captain Cook's death reached England

another early friend of Franklin's, Lord Howe, on hearing of the passport incident, told the King that he would like to give Franklin a copy of Cook's *Voyage to the Pacific Ocean* on behalf of the British Admiralty. George III gave only grudging consent, for he then had feelings of bitterness against Franklin (page 80).

The Royal Society, whose President Sir Joseph Banks had sailed with Cook on a previous voyage of exploration, decided to strike a medal to commemorate Cook's voyages, and it was agreed to cast a few of the medals in gold. The Society resolved that one should be presented 'to the rebel Franklin'. This Sir Joseph dispatched with a note that the medal was presented 'in testimony how truly they respected those liberal sentiments of general philosophy which induced you to issue your orders to such American cruisers as were then under your direction to abstain from molesting that great navigator'.[3]

*   *   *   *   *

Napoleon, Emperor of France, was England's most bitter enemy and it is universally recognised that he knew the art and practice of war as well as any of the world's greatest generals. It might, therefore, have been expected that he would have frowned on any attempt at fraternisation or friendliness between civilians of enemy countries. But such was not always his attitude.

He was greatly interested in medical science and any new discovery likely to improve the health of his people received his close attention. An outstanding example of this attitude was the manner in which he quickly decided that the method of vaccination discovered by Jenner would be of immense value to his nation. He showed his faith in the newly discovered method by having his own young child vaccinated, and, in the year 1809 he sanctioned a national decree ordering vaccination.

In the year 1804, only one year after the war with England had started, one of the most beautiful of the Napoleonic series of medals was struck. This commemorated the Emperor's estimation of the value of vaccination. It has been said that it was at the same time intended as a mark of personal honour to Jenner. Jenner's biographer, Baron, had this to say of Napoleon:

45. *Courtesies to Enemy Scientists*

He who was flushed with victory, and at the head of the revolutionary army of France, had spared the University of Pavia, out of respect of the genius of Spallanzani when the city itself was given up to plunder, proved that the claims of science were not forgotten amid the astonishing events which carried him forward to the highest pinnacle of ambition.

For it happened that two Englishmen were interned in France, and Jenner petitioned for their release. Napoleon, it is said, was about to reject the petition when his Empress, Josephine, uttered the name of Jenner. The Emperor paused for an instant and exclaimed, 'Jenner. Ah we can refuse nothing to that man', and the two men were given their liberty.

*   *   *   *   *

Napoleon also honoured another British scientist during wartime. Shorly after Volta's invention of a chemical method of producing electricity (page 86), Napoleon offered a medal and three thousand francs to be awarded annually for the best experimental work on electricity, In 1807 these were awarded to Sir Humphry Davy although England and France were then at war. One writer commented:

Thus did the Voltaic battery, in the hands of the English chemist, achieve what all the artillery of Britain could never have produced,– a spontaneous and willing homage to British superiority.

Davy wrote:

Some people say I ought not to accept this prize; and there have been foolish paragraphs in the papers to that effect; but if the two countries or governments are at war, the men of science are not. That would, indeed, be a civil war of the worse description: we would rather, through the instrumentality of men of science, soften the asperities of national hostility.[4]

Later, Davy was elected a corresponding member of the First Class of the Imperial Institute and visited France during the war. It was related that:

Nothing ever exceeded the liberality and unaffected kindness and attention with which the learned men of France received and caressed the English philosopher. Their conduct was the triumph of science over national animosity,– a homage of genius, alike honourable to those who bestowed and to him who received it.

Sir Humphry was even invited to an anniversary dinner where complimentary toasts were proposed to the Royal Society of London and to the Linnean Society of London: and all this was during the war against England. At the dinner, we are told,

the circumstances which evinced the greatest feeling and delicacy towards their English guest, was the company's declining to drink the health of the Emperor. It placed their personal safety in some jeopardy; and not a little apprehension was afterwards felt as to how far Napoleon might resent such a mark of disrespect.

But Napoleon took no action, so all was well.

46. *Kings, Rulers and Scientists*

SHORTLY AFTER the restoration of Charles II to the throne of his father much interest was aroused in popular science, largely through the activities of a body of men called 'the Royal Society of London for Improving Natural Knowledge'. The King was keenly interested in many scientific matters and gladly became patron of the Society; and many courtiers and others followed his example. Before long its members included 'persons of the degree of baron, fellows of the College of Physicians and also professors of mathematics, physics, and natural philosophy in the Universities of Oxford and Cambridge'. Lord Macaulay, the well-known historian, made this comment about the scientific interests of that time.[1]

Within a few months of the foundation of the Royal Society experimental science became all the mode. Dreams of perfect forms of government made way for dreams of wings with which men were to fly from the Tower to the Abbey, . . . Cavalier and Roundhead, Churchman and Puritan, were for once united. Divines, jurists, statesmen, nobles, and princes swelled the triumph. Poets sang with fervour the approach of the Golden Age. Dryden predicted that the Royal Society would soon lead us to the extreme edge of the globe, and there delight us with a better view of the moon. Charles himself had a laboratory at Whitehall and was far more active and attentive there than at Council boards. It was almost necessary to the character of a fine gentleman to have something to say about air pumps and telescopes; and even fine ladies, now and then, thought it becoming to affect a taste for science, and broke into cries of delight at finding that a magnet really attracted a needle and that a microscope really made a fly look as large as a sparrow.

Rupert, Prince of Bavaria, was one of the most enthusiastic members of the Society. He was a nephew of Charles I, in whose army he had served during the Civil War. After his uncle's defeat he went to the continent, where he remained until the Restoration. Whilst in exile he spent much time in scientific studies. He returned to England with his cousin Charles II. His researches and discoveries in science merit him a place among the famous royal scientists of the world.

One of his most important pieces of research was on gunpowder, a not surprising subject for a former warrior. It is reputed that he made a powder which had ten times the ordinary strength of that used in his day. Other discoveries of his were a method of blowing up rocks in mines or under water; a hydraulic engine; a method of making 'hail shot'; an improvement in the naval quadrant; and an improvement in the locks of firearms. Amongst his chemical discoveries were the composition of an alloy now called Prince's metal and a method of rendering black lead fusible.

One invention which has been of great interest to many people is that known as Prince Rupert's Drop.[2] Macaulay spoke of it as 'that curious bubble of glass which has long amused children and puzzled philosophers'.

This puzzle was introduced by Rupert into England in 1660, and communicated by Charles II to the Royal Society at Gresham College.

The glass bubble is a solid piece of glass, somewhat pear-shaped, like a tadpole with a long tail. It is formed by pouring highly refined molten glass into cold water. Its thick end, or 'head', is so hard that it can scarcely be broken on an anvil. But the taper end of its tail is easily broken off. When this is done, the whole piece disappears as dust, with a sharp explosion; it may do so even when the tail is scratched or when it is cut to a certain depth. It should be added that though they seem simple in structure, these drops are difficult to make.

Rupert's drops were very well known when Butler wrote these lines in his famous *Hudibras*:

> Honour is like that glass bubble
> That finds philosophers such trouble,
> Whose least part cracked, the whole does fly
> And wits are crack'd to find out why.

*   *   *   *   *

Napoleon, like Charles II, took a personal interest in his own country's leading scientific society, the French Académie Royale des Sciences, which had been founded in the year 1666. He also appreciated the importance of science in warfare and took scientists with him on his expedition to the Nile. One story about Napoleon is well worth repeating since it shows that even rulers cannot order science to their will, as George III had learned only a few years before (see chapter 36).

Napoleon was attending a gathering of scientists and was told of Davy's success in obtaining the metal sodium by the use of electricity. He asked why such a discovery had not been made in

Napoleon gets an electric shock

France. 'We have never constructed a voltaic battery of sufficient power', the scientists answered. 'Then', exclaimed Napoleon, 'let one be instantly formed without any regard to cost or labour'.

One was made and Napoleon went to see it. He was told of the peculiar taste produced when the ends of both wires that are joined to the poles are put to the tongue. Then, according to one account:

With that rapidity which characterised all his motions, and before the attendants could interpose with any precautions, he thrust the two ends of the wires of the battery under his tongue and received a shock which nearly deprived him of all feeling. After recovering from its effects, he quitted the laboratory without making any remark, and was never afterwards heard to refer to the subject.[3]

*   *   *   *   *

Many years before this incident of Napoleon and the electric battery, Berthollet, a celebrated French scientist, ran the very great risk of being guillotined for refusing to obey the order of Robespierre, who, at that time, held the power of life and death in Republican France.[4]

Claude Louis Berthollet was appointed physician to the Duke of Orleans in 1772 and later became the director of the French Government dye-works. He had won for himself a high scientific reputation by the time the French Revolution burst upon the world. Then all the great powers of Europe combined to attack France, and the Austrian and Prussian armies hemmed her in by land, while the British fleets blockaded her by sea. France was thrown at once on her own resources. She had been in the habit of importing saltpetre (to make gunpowder) as well as iron and many other things necessary for war. These supplies being suddenly cut off, it seemed likely that France would have to submit to any terms imposed upon her by her enemies.

The Republican leaders called their men of science to their assistance. One who answered the call was Berthollet and after much experimentation he showed how saltpetre could be made from French soil. It has been said that he also invented a method of smelting iron and then of converting it into steel. The importance of his work can be judged from this comment: 'It was in all probability his zeal, activity, sagacity, and honesty, which saved France from being overrun by foreign troops.'

As is well known, during the Reign of Terror the Republican leaders pretended to discover plots as an excuse for putting to death those whom they wished to destroy, and Robespierre, the Republic's leader at the height of the Reign of Terror, hatched this plot to enable him to convict many of his enemies. He gave notice at a sitting of the Committee of Public Safety of the discovery of a conspiracy to put many soldiers to death. The plot, he said, was to poison the brandy, a stimulant usually given to soldiers just before they went into battle and also to sick soldiers in hospital. He added that some of those in hospital who had tasted this brandy had in fact been poisoned by it.

The Committee gave immediate orders to arrest the men named, all of whom had been previously ear-marked by Robespierre for execution. To get the necessary evidence for the trial a quantity of the brandy was sent to Berthollet for analysis. He was informed, at the same time, that Robespierre himself had asked for this evidence to convict his enemies and that anyone disobeying him would meet with certain destruction. Having finished his analysis, Berthollet sent his report to the republican leaders. He stated, in the plainest language, that nothing poisonous was mixed with the brandy, but that it had been diluted with water which contained small particles of slate in suspension, an ingredient which filtration would remove.

This report obviously upset the plans of the Committee of Public Safety. They sent for the author and tried to convince him of the inaccuracy of his analysis and so make him alter his report. Finding that he remained unshaken in his opinion, Robespierre exclaimed 'What, Sir. Darest thou affirm that the muddy brandy is free from poison?' Berthollet immediately filtered a glass of it in his presence, and drank it off. 'Thou art daring, Sir, to drink that liquor', exclaimed the ferocious President of the Committee. 'I dared much more', replied Berthollet, 'when I signed my name to that Report'.

46. Kings, Rulers and Scientists

Berthollet drinks the brandy

There can be no doubt that he would have paid the penalty with his life for this undaunted honesty but that the Committee of Public Safety could not at that time dispense with his services.

When Napoleon came to power he recognised the great abilities of Berthollet and showered honours upon him, later raising him to the peerage with the rank of Count.

*　　*　　*　　*　　*

Queen Victoria's husband, the Prince Consort, was a firm disciplinarian. His attitude towards the education of boys was well summed up by his own phrase, 'Never relax'. The young Prince of Wales, the future Edward VII, was brought up on these strict principles.

All arrangements had been made for the Prince to go to Oxford University in the October of 1859 when he returned from a foreign tour. It so happened that he returned to England earlier than expected – in July instead of October. He had therefore three months to spend before going to Oxford. But he was not allowed to have this period for a holiday. Instead, his father sent him to Edinburgh University where the time could be spent profitably, in study. A suitable course was mapped out for him, which included lectures on chemistry by Dr Lyon Playfair. These lectures were illustrated by practical work and visits to various industries.

At a lecture one day, when the Prince was sitting on the same bench as the sons of Scottish peers and peasants, Dr Playfair wished to explain why Algerian conjurers could apply hot irons to their bodies without injuring themselves. This, he said, was possible if the metal was raised to a sufficiently high temperature. He had close at hand a cauldron of lead which was boiling at white heat, its temperature being in the neighbourhood of 1500°C to 1700°C.

The professor suddenly turned to the Prince and said, 'Now, Sir, if you have faith in science you will plunge your right hand into that cauldron of boiling lead and ladle it out into the cold water standing by'. 'Are you serious?', asked the Prince. 'Perfectly', was the reply. 'If you tell me to do it, I will', said the Prince. Thereupon the professor carefully washed the Prince's hand with ammonia, to dissolve any grease or natural oil that might be on it. Then the Prince put his hand in the boiling lead and ladled some of it out without injuring his hand in the slightest.[5]

The incident illustrates first of all the ready obedience with which the Prince carried out an order and which justified, to some extent, the strictness of his upbringing. It also shows his great courage, for very few people would be brave enough to do as he did even at the bidding of such a distinguished scientist as Playfair.

The incident also illustrates, in a dramatic manner, the scientific fact that the hand, when *perfectly* clean and free from grease, can be placed in molten lead without injury. This is possible because the 'natural' moisture of the skin acts as a kind of cushion between the lead and the skin. The lead forms small globules which bounce off the hand as do the globules of mercury when the hand is placed in a dish of this liquid.

Before concluding this account it is necessary to give a strict warning to any young scientist who is considering doing this experiment. For it must be emphasised most strongly that an inexperienced person can suffer great injury unless he knows how to prepare the hand before it is plunged in the molten liquid, and no one should attempt the experiment without expert supervision.

47. Two Mathematical Problems of Old

FIVE HUNDRED years before the birth of Christ an Italian named Zeno left his native land and went to Greece, where he studied under one of the wise men, or philosophers. Later in life, Zeno, who is thought to have been somewhat of a wit, asked the mathematicians of his day, four awkward questions.[1] One of these had to do with racing, 'Why is it', said he, 'that the faster of two runners can never overtake the slower, if the slower is given any start at all? For', continued Zeno, 'the slower will have left a given spot when the faster man reaches it, and therefore the slower runner must always be ahead'.

This question was put in a simpler way as an imaginary race between Achilles and a tortoise. Achilles, the Greek god, was reputed to be the swiftest runner ever known (legend has it that even at the age of six he could overtake a running stag). The tortoise is, of course, one of the slowest moving of all creatures.

Here, then is the centuries-old problem. Suppose, for example, that Achilles runs ten times as fast as the tortoise and gives it 1000 yards start; when will he overtake it? The argument is as follows: When Achilles has run 1,000 yards the tortoise will have crawled 100 yards, and so will, of course, be 100 yards in front. By the time Achilles has run this 100 yards the tortoise will have crawled 10 yards and will be 10 yards in front. By the time

Achilles has run those 10 yards the tortoise will have crawled 1 yard, and will still be a yard ahead. By the time Achilles has run this yard the tortoise will be ahead by 1/10 of a yard. Again, when Achilles has run this 1/10 of a yard, the tortoise will be 1/100 of a yard in front – and so on, *ad infinitum*. 'Where', asked Zeno, 'is this argument wrong?'

Mathematically, it certainly seems that Achilles can only keep getting nearer and nearer to the tortoise, but can never overtake it. At one time, by mathematical reckoning, he would be within 1 yard of it, a little later he would be only say 1/1000 of yard behind it, and so on. But he would always be behind, even though by only a tiny fraction of a yard.

Much has been written in the last two thousand years about this problem and many mathematical solutions have been suggested. If the problem has served no other purpose it has clearly shown the care which must be used when everyday problems are treated as mathematical exercises.

The mathematicians have dealt with the race as if it were made up of a number of short races, the first being of 1,000 yards, the next of 100 yards and so on, the separate races being 1,000, 100, 10, 1, 0·1, 0·01, 0·001, 0·0001, 0·00001 yards – and so on.

It will be seen that these distances become smaller and smaller until they become infinitely small. The problem, therefore, has to do mathematically, with these infinitely small numbers.

It would probably be wise to avoid following the example of the German professor who once tried explaining the problem. The Queen, to whom he was speaking, soon grew tired of his learned explanation, and told him that she knew all there was to know about the infinitely small, because for many years she had had to deal with courtiers and statesmen!

The only explanation which will be given here is that the race is not made up of a number of very, very tiny distances but is continuous until the end.

In any case it is often best to call commonsense to our aid when dealing with mathematical and statistical problems. This, of course, tells us from our day-to-day experience that the swifter runner would soon pass the slower one.

47. *Two Mathematical Problems of Old*

It is generally agreed that geometry was first widely practised in Egypt – that is, the land bordering on the River Nile. This river often overflows its banks and when it does the mud carried down from the high ground is deposited on the fields near the delta. This flooding, and the consequent depositing of mud, blots out the landmarks. Therefore after each flood the Egyptians had to refix the boundaries of the fields. For this reason the early Egyptians learned how to measure fields bounded by straight lines and to calculate the areas of the space enclosed.

In course of time philosophers became very interested in straight lines and the figures enclosed by them, as well as in curves and circles and so forth. Hence the study of geometry became theoretical as well as practical, and the philosophers found a great interest in trying to solve problems by geometrical methods, using only a straight edge or ruler and the compasses.

One such problem which they never solved by these methods was that of drawing a cube which is *exactly* twice as big as a given cube, a problem known to mathematicians as that of duplicating the cube.

One of the different legends about its origin relates that Minos,[2] the legendary King of Crete, had a young son named Glaucus who some say was playing at ball and others say was chasing a mouse when he fell into a tub of honey and was smothered. The King called his soothsayer and cruelly ordered him to restore the child to life or be buried alive with the body of the boy. The soothsayer, being unable to raise the boy, was accordingly buried with him. But whilst in the tomb he found a way to awaken the boy, who was apparently only in a trance, and then restored him to his father.

The story does not mention whether this incident led Minos to think of his own death, but the legend has it that he ordered a tomb to be erected for himself as soon as his boy was 'raised' from the dead. He gave orders that the tomb must be in the shape of a cube. The King went to inspect the finished tomb and expressed his grave displeasure that the builders considered such a small tomb suitable for a royal prince. He ordered them to make one twice as great!

But doubling the length of the sides the builders would have built a tomb whose volume would have been eight times that of the original one and not 'twice as great' as the King had ordered.

So they consulted the wise men of the day as to how they should design a cube *exactly* twice the volume of the original one. The wise men failed to give them a satisfactory solution. Every mathematician since then who has attempted to solve this problem using only the methods of geometry already described has also failed.

In the other legend the problem of duplicating the cube originated at Delphi, a town of ancient Greece.[3] There the temple was dedicated to the Greek God named Apollo. The Greeks believed that their many Gods had great powers and that almost everything that happened was governed by one or other of them. For example, they thought that the God Apollo had the power of sending plagues and epidemics to punish man. They believed, also, that he could make them free from these illnesses if he wished.

It so happened that a serious epidemic broke out and the people thought that Apollo had sent the epidemic because he was displeased with them. They therefore decided to send their leaders to his temple to plead that they be delivered from the plague.

Each God had his own temple which was in the charge of a priest or a priestess, who spoke to the God on behalf of the people and conveyed his orders to them; for the God was thought to be much too important a being to speak to ordinary people. The place at which the God conversed with his people, through the mouth of his priest, was called the oracle.

According to the legend, the leaders of the Greeks went to the Temple of Apollo at Delphi, where the priestess was a holy woman, and asked her to plead with the God to remove the plague. The priestess did so; and Apollo 'through the oracle' told the people that he would take away the plague on condition that they built an altar similar in shape but twice as big as the existing one.

The existing altar was in the shape of a cube and no one knew how to build one exactly twice as great. So in despair they consulted Plato, the wisest man of Greece. According to one legend he told them that Apollo had not meant them to attempt a task requiring the highest skill in geometry, neither did the God specially want them to build an altar double the size of the present one; he had spoken to them through the oracle to let them know that he wished them to study geometry much harder than they had done previously!

48. Members of Parliament were not Mathematicians

NO MATHEMATICAL test is, of course, given to would-be members of parliament, but a sum was once set in the House of Commons which, it was said, only a few members could do.

In March 1866 Gladstone, then Chancellor of the Exchequer, introduced a Reform Bill which has since been termed a moderate one but which, at the time, led to a serious parliamentary crisis. It was designed to give the vote (or in parliamentary language, to extend the franchise) to 'hundreds of thousands of his fellow countrymen'.

One of the items of discussion during its passage through the House was whether the vote should be given only to persons of 'decent education', for example, to a man who could pass a simple test in dictation.* This was opposed by Gladstone. In moving its rejection he stated that one of his objections to any educational test was 'that as the life of labour was not favourable to the retention of scholastic education, a man might possess the franchise long after he has lost the qualification for it'.

He continued:

At present our electoral system is totally devoid of any educational test and no one finds that there is any serious necessity for such a test to the working

---

* It should be mentioned that in 1866 thousands of adults had had little or no education and many could neither read nor write.

of our system. If some test, simple in its character could be devised, some test capable of being applied to voters of all classes without raising odious distinctions between men in different walks of life, it might perhaps be wise to accept such a test. The nearest test of that kind that suggests itself to my mind is that the voter should be asked to sign his own name.

After dealing with some objections to this and other proposals, he continued:

What is writing from dictation? It is a most severe trial and one in which failure is the lot of myriads of young men who apply not for offices of manual labour but for clerkships. Yet the House of Commons is asked to make satisfactory writing from dictation the condition of attaining the electoral franchise.

The debate then continued:

Putting aside subtraction and multiplication, I should like to know how many of the labouring classes can pass an examination in division of money or how many members of this House can pass such an examination.[1] If I give the sum £1,330 17s 6d and tell members of this House to divide it by £2 13s 8d I want to know how many would do it.

Mr Hunt: 658. [This was the number of members of Parliament.]

The Chancellor of the Exchequer: There are not three or four in this House who could do it. I would say there are not thirty or forty, without the least fear of contradiction. I will go further and say it is not necessary that they should; and that they may be admirable members of this House without being able to work such a sum.

Lord R. Montague: You cannot divide by £2 13s 8d. (Laughter).

The Chancellor of the Exchequer: One illustration is better than a thousand arguments. The noble Lord is one of the more promising financial members of the House and he tells us positively that division of money is a thing that cannot be done.

Later in the proceedings Lord Robert Montague enlarged on his former interjection, thus:

With regard to the sum of division which the Right Honourable Gentleman has suggested, it was quite possible to divide the sum of money, but not by *money*. How could one divide money by £2 13s 8d? The question might be asked, 'How many times 2s will go into £1, but that was not dividing by money; it was simply dividing 20 by 2. He might be asked, 'How many times will 6s 8d go into a pound, but that was merely dividing 240 by 80.

Money, according to Lord Robert, cannot be divided by *money*; it can only be divided by a *number*.

154

In the previous chapter mention was made of problems which puzzled the ancient Greeks. Another of these problems is the one known as squaring the circle, that is, of drawing a square which is exactly the same area as a given circle. Many attempts have been made to do this, both by Greek mathematicians and by others of more recent times; but none has been successful when restricted to the use of only a straight edge and compasses.

Incidentally, no one has yet succeeded under the same restrictions of using only the ruler and compasses in drawing a straight line of equal length to the circumference of a given circle. The length of such a circumference is now denoted by mathematicians by the formula $2\pi r$ where 'r' is the radius of the particular circle, and π (a Greek letter pronounced pie) stands for a number which is the same for all circles and is usually given as 3·14. But its correct value has never been found. One mathematician calculated its value to over thirty decimal places and, his value of π, with the thirty or more figures after the decimal point, was chiselled on his tomb-stone! Other mathematicians have made similar calculations, so that now π can be given to over 700 figures after the decimal point. But even though it can be determined to 700 decimal places the *exact* value of π has not been found.

An interesting story about squaring the circle comes from Cambridge University – a university long famous for its mathematical learning. Mathematicians have their own terms, and 'squaring' the circle is to them the 'quadrature' or the 'rectification' of the circle.

In this story another mathematical problem is touched on. Most readers at one time or another will have seen a curved light on the surface of a liquid contained in a cup. This curve is due to reflection of light from the curved surface of the cup and is known as a 'caustic'. The mathematical study of the 'caustic curve' is a very difficult one.

The story runs as follows: Henry Goulburn, after leaving Trinity College, Cambridge, became a member of Parliament and in 1826 was appointed Chancellor of the Exchequer. Five years later he offered himself as a parliamentary candidate for the

University, which at that time sent two members to the House. Some years before, Goulburn, in his capacity as Chancellor, had displeased some of the mathematicians by telling a deputation from the Astronomical Society that the Government 'did not care twopence for all the science in the country'.

The election was in May 1831, and Lord Palmerston and Mr Cavendish, two Whigs, opposed the Right Honourable Henry Goulburn and Mr W. J. Peel, two Tories. Feeling ran high, for electioneering was an exciting affair even in those days.

Late one evening in early May, a cab drove up in hot haste to the office of the *Morning Post* – a daily paper which was strongly Tory. A man got out and handed in an advertisement stating that it was from Mr Goulburn's Committee. At the same time he ordered fifty extra copies of the next morning's *Post* to be sent to Goulburn's Committee room.

Next morning, on 4 May, the paper appeared and one paragraph read as follows:

We understand that, although owing to circumstances with which the public are not concerned, Mr Goulburn declined becoming a candidate for University honours, his scientific attainments are far from inconsiderable. He is well known to be the author of an essay in the *Philosophical Transactions* on the accurate rectification of a circular arc, and of an investigation of the equation of a lunar caustic – a problem likely to become of great use in nautical astronomy.

Although the appearance of this caused tremendous amusement among the mathematical dons of the University – at any rate amongst those with Whig views – it evidently did no great harm to Mr Goulburn's chances, for he was elected. But the advertisement soon took its place in the history of the many hoaxes which have been perpetrated in the University.

It was, indeed, a clever bit of work, for there was much in it which could have been true. Thus, anyone who had solved the problem of squaring the circle would unquestionably have been given the highest University honours possible. Moreover, the *Philosophical Transactions*, a paper published by the Royal Society, contained only those 'essays' which are of the highest merit, mathematically or scientifically.

The advertisement was cleverly worded so that the editor of

the *Post*, on his hasty reading of a note handed in at the last minute, could be excused from not questioning the term'rectification of a circular arc'. For this term was not often used whereas if the hoaxer had used the customary phrase of 'squaring the circle', the editor would most probably have spotted the hoax. The term 'lunar caustic' was also well chosen. For the word 'lunar' means, of course, the moon, and it was quite possible that some brilliant mathematician might have investigated something which had to do with the moon.

But the mathematicians and scientists in Cambridge chuckled greatly over this 'equation of a lunar caustic'. They had often studied 'caustic curves', but 'lunar caustic' was another name for silver nitrate (a substance which was cast into small white sticks and used in surgery to cauterise, or burn, a wound).

The person suspected of being responsible for the hoax was Charles Babbage, the well-known mathematical don (chapter 16). In his famous book, *Passages*, he recalls the hoax, in these words. 'I remember a very harmless squib which I believe equally amused both parties, and which, I was subsequently informed, was concocted in Mr Cavendish's committee room'. He then gives all the circumstances in such detail that another Cambridge mathematician wrote, 'I think the man – the only one I have ever heard of – who knew all about the cab and the extra copies must have known more' – meaning, of course, that he thought that Babbage was the originator of the hoax.

Mr Goulburn, as we have already mentioned, was successful at the poll and represented the University in Parliament until his death. In the 1830's a parliamentary election was almost a 'free fight' for all, with little or nothing barred. Today a hoax like this could result in extremely serious penalties being inflicted on the persons convicted of being responsible for it.

A most interesting conclusion to the episode is that only four years after the hoax Goulburn's son was declared the Second Wrangler, a title given to the undergraduate who came second in the list of successful candidates in the Mathematical Tripos. The Tripos is the final examination of the University for the degree of Bachelor of Arts.

49. The Scientists are Taught to be Cautious

CHARLES II, King of England, was greatly interested in the science of his day and took particular pleasure, we are told, 'in experiments relating to navigation, of which he had a very accurate knowledge, and paid great attention to finding out what sorts of wood required the least depths of water to float them and what shapes are best adapted for cutting water and making good sailors'. This royal interest in floating objects adds point to the following story:

One day, when the Fellows of the Royal Society (page 142) were met, Charles put to them this problem: 'If you first weigh a basin containing water and again weigh it with a live fish in it, the weight remains the same. But if you weigh a basin containing water and again weigh it after putting a dead fish in it, the weight increases by an amount equal to the weight of the dead fish'.[1] He asked the reason for this.

Most of the Fellows of the Society knew about Archimedes's investigation of the goldsmith's crown (chapter 24), and also that a solid object when suspended in water weighs less than it does in air. But none could answer this question immediately. Yet the King had set a problem, and it must be answered – the prestige of the Society was at stake.

The writings of Archimedes and those of other scientists were consulted and a long discussion followed. Many 'learned' reasons were given to explain this previously unheard-of problem, but none was satisfactory.

After they had spent some considerable time in discussing the problem one of the Fellows remembered a good rule which he had been taught at school. It was given in a Latin phrase: 'consider first the *an sit* of a statement before investigating the *cur sit*'. This means that before discussing *why* a certain thing happens, it is wise, first of all, to make sure that it *does* really happen. So, with great daring, he suggested that before considering the reasons why there should be a difference in weight between a live fish and a dead one in the bowl of water, they ought first of all to find out whether there really was such a difference.

This daring statement was too much for the assembled scientists and courtiers. They were horrified that anyone should question the King – he could do no wrong! One Fellow declared that it was an act of treason even to doubt the King's statement, but to say openly that the King had made a false assertion was very dreadful indeed. Others jumped up to say that they had known for many years that what the King had said was perfectly true – that it was a fact that a live fish caused no increase in weight when put into water but that a dead one did.

A long time was spent in fruitless discussion, and then the Fellow suggested again that they should see for themselves what really did happen. This time a basin of water was fetched and

Charles II puzzles the scientists

then weighed. Into it was put a live fish, and, with bated breath, all awaited the result whilst the basin and fish were weighed. This weight was greater than that of the basin of water.

The live fish was taken out and left to die. Then it was put into the basin and another weighing was made. The basin of water with the dead fish in it weighed exactly the same as with the live fish in it. They then realised that Charles, the well-named Merry Monarch, had played one of his pranks on them.

*   *   *   *   *

The story makes good reading and has a moral which all might well remember; but it is probably a fictitious tale. No mention of this incident can be found in any of the reliable histories of the Royal Society, although it is almost certain that such a royal joke would have been mentioned had it really taken place. On the other hand, history does record that the Royal Society at one period did become the object of scorn, especially from one man who was refused membership. In revenge he invented all kinds of silly stories about the Society, none of which was true. One such story, a most amusing one, was based on a book which had then just been written recommending 'tar water' as a medicine to 'keep the blood in good order'.

At one of its meetings, the Royal Society received a communication addressed from Portsmouth stating that a sailor had broken his leg in a fall from a mast-head but that bandages and 'a plentiful application of tar-water' had enabled him, in three days, to use his leg as well as ever. This communication, according to the story, had been under grave discussion for a long time when another letter was delivered in haste, by hand at the door of the room. It read that the writer had forgotten to mention in his first letter that the sailor's leg was a wooden one![2]

The story of Charles and the live and dead fish is probably on a par with this and may well have been invented by the same man.

There is, however, another way in which the fish story could have started. As so often happens, a very similar story had been told years before; actually in 1660. It is said that Louis XIII, King of France, asked his courtiers to consider why it was that when a live fish was thrown into a bowl full of water some of the water overflowed, but when a dead fish was thrown in, none flowed out. The courtiers considered and considered but could find no reason. At last they sent for the gardener to bring a bowl and a live fish. He filled the bowl to the brim with water and threw in the live fish. The water overflowed. He then took out the fish and when it had died put it back in the refilled bowl. They saw that the water overflowed again.[3]

One version of the story about Charles II reports that after the scientists had tired themselves out with arguing one of the men boldly said that the statement was false and needed no argument, whereupon the King, in high mirth, exclaimed: 'Odds*fish*, brother, you are right'.[4]

References, Chapters 23-27

CHAPTER 23 (pp. 9–12)

1. Vitruvius, *De Architectura*, vol. IX, pp. 1–9.

CHAPTER 24 (pp. 12–19)

1. Plutarch, *Lives, Marcellus*, vol. XIV.
2. *Ibid.*
3. Polybius, *History*, vol. VIII, 7–8.
4. Plutarch, *Marcellus*, vol. XV.
5. *Ibid.*, vol. XIX.
6. Livy, vol. XXV, 31.
7. Aristophanes, *The Clouds*, pp. 765–70.
8. *Phil. Trans.* abridged, vol. VII, pp. 345, 558.
9. W. W. Rouse Ball, *A Short History of Mathematics* (1920), p. 65.

CHAPTER 25 (pp. 19–26)

1. *The Dictionary of Mr Peter Bayle*, ed. Des Maizeaux, 2nd ed. 1734–38.
2. Pliny, vol. XXXIV, 14.
3. Vincent Le Blanc, *The World Surveyed* (1660), p. 13. Also *The Spectator*, Number 191.
4. R. F. Burton, *Pilgrimage to Al-Madinah and Mecca* (1893), vol. II, p. 339.
5. W. Irving, *The Life of Mahomet* (1915 ed.), p. 228.
6. Bayle's *Dictionary*.
7. W. Barlow, *Magneticall Advertisements, etc.* (1616), pp. 35–6.
8. Bayle's *Dictionary*.
9. *Encyclopaedia Britannica*, 9th ed., vol. XIII, p. 274.
10. *The Thousand and One Nights*, trans. by Wm. Lane (1839), pp. 179 ff.

CHAPTER 26 (pp. 26–32)

1. W. Irvin, *Life and Voyages of Christopher Columbus* (1890), p. 77.
2. G. Duff, *The Truth About Columbus* (1936), p. 124.
3. A. C. Miller, *History of Terrestrial Magnetism and Atmospheric Electricity*, vol. XLII, (1937), pp. 268 ff.
4. A. C. Miller, *op. cit.*, p. 268.
5. A. Larsen, *The Discovery of Electromagnetism by H. C. Oersted* (1920), p. 39.
6. H. Bence Jones, *Life and Letters of Faraday* (1870), vol. II, p. 390.

CHAPTER 27 (pp. 32–8)

1. K. von Gebler, *Galileo Galilei and the Roman Curia*, trans. Mrs G. Sturge (1879), p. 10.

References, Chapters 27-32

2. J. J. Fahie, *Galileo, His Life and Work* (1903), p. 24. Also R. A. Gregory, *Discovery* (1917), p. 2. Also I. V. Hart, *Makers of Science* (1923), pp. 105–6.
3. Lane Cooper, *Aristotle, Galileo and the Tower of Pisa* (1935), p. 14.
4. E. J. Dijksterhuis, *The Principal Works of Simon Stevin* (1955), p. 551.

CHAPTER 28 (pp. 38–44)

1. J. Priestley, *The History and Present State of Discoveries Relating to Vision, Light and Colours* (1772), p. 86.
2. Galileo Galilei, *The Sidereal Messenger*, trans. E. S. Cripps (1880), pp. 9–11.
3. K. von Gebler, *op. cit.*, pp. 18–19.
4. *Ibid.*, p. 6.
5. J. J. Fahie, *op. cit.*, p. 9.
6. *Chambers Encyclopaedia* (1955), vol. VII, p. 220.

CHAPTER 29 (pp. 45–51)

1. Ecclesiastes, vol. I, p. 5.
2. Joshua, vol. X, pp. 12–13.
3. All the quotations on the trial are from *The Gentleman's Magazine*, vol. XV, pp. 584 ff.
4. Psalms 6, 32, 38, 51, 102, 130 and 140.
5. Guiseppi Baretti, *The Italian Library* (1757), p. 52.
6. *Nature*, vol XXXVII (1936), p. 10. Also J. J. Fahie, *Memorials of Galilei Galileo 1564–1642* (1929), pp. 74–5.

CHAPTER 30 (pp. 51–6)

1. Supplement to *Encyclopaedia Britannica*, 4th, 5th and 6th eds. (1824).
2. Principal Tullock, *Pascall* (1878), pp. 34–6.

CHAPTER 31 (pp. 57–60)

1. Supplement to *Encyclopaedia Britannica, ed. cit.*
2. Otto von Guericke, *Experimenta nova, ut vocant, Magdeburgica de vacuo spatio*, from Oswald's *Klassiker* (1894), no. 59.

CHAPTER 32 (pp. 61–5)

1. Edmund Turnor, *The Town and Soke of Grantham* (1806), p. 160.
2. R. Greene, *The Principles of the Philosophy of the Expansive and Contracting Forces* (1727), Appendix.
3. R. Voltaire, *Letters Concerning the English Nation* (1741), p. 127.

4. H. Pemberton, *Sir Isaac Newton's Philosophy* (1728), Preface.
5. *Notes and Queries*, Second Series, vol. v, p. 312.
6. W. Stukeley, *Memoirs of Sir Isaac Newton's Life*, ed. by A. H. White, (1936), pp. 19–20.
7. D. Brewster, *Memoirs of the Life, Writings and Discoveries of Sir Isaac Newton* (1855), vol. II, p. 416.
8. A. de Morgan, *Newton* (1885), p. 14.
9. Maude, *Wensleydale* (1787). Also M. Biot, *The Life of Newton* (1852), p. 25.
   (In the *Biographie Universelle*).
10. W. Stukeley, *op. cit.*, p. 60.

CHAPTER 33 (pp. 66–73)

1. *Phil. Trans.* Abridged ed., vol. VIII, pp. 2 ff. and vol. VII, p. 499.
2. J. A. Nollet, *Essai sur l'Electricite Des Corps*, (1746).
3. *Phil. Trans.* Abridged, vol. IX, p. 265.
4. *Ibid.*
5. *Phil. Trans.*, vol. XLIII, p. 481. Also *Phil. Trans.*, vol. XLIV, pp. 41, 388 695, 704.

CHAPTER 34 (pp. 73–81)

1. *Life of Dr Franklin* published by T. Kinnersley (1816), p. 95.
2. J. Sparks, *The Works of Benjamin Franklin* (1840), vol. v, pp. 288–9.
3. B. Franklin, *Experiments and Observations on Electricity, etc.* (1796), pp. 111–12.
4. J. Priestley, *The History and Present State of Electricity, etc.* (1767), pp. 117–21.
5. B. Franklin, *Poor Richard's Almanack for 1753*. (In I. B. Cohen, *Benjamin Franklin's Experiments* (1941), p. 129.)
6. *Phil. Trans.* Abridged, vol. x, p. 527.
7. C. R. Weld, *History of the Royal Society* (1847), vol. II, p. 100.
8. I. D. Israeli, *Curiosities of Literature* (1824), p. 528.

CHAPTER 35 (pp. 81–6)

1. *The Gentleman's Magazine*, May, 1799. Also S. Muspratt, *Chemistry* (1860), vol. I, p. 782.
2. A. Galvani, *De viribus electricitatis in motu musculari* in **Oswald's** *Klassiker*, Number 52, pp. 4–5.
3. *Encylopaedia Britannica*, 9th ed., vol x, p. 49.

CHAPTER 36 (pp. 87–92)

1. James Raine, *A Memoir of Rev. J. Hodson* (1857), vol. I, ch. IX.
2. S. Smiles, *Lives of the Engineers – G. and R. Stephenson* (1879), p. 95.

References, Chapters 36-41

3. J. A. Paris, *Life of Sir Humphry Davy* (1831), pp. 119, 131.
4. S. Smiles, *op. cit.*, p. 104.
5. S. Smiles, *op. cit.*, p. 108.

CHAPTER 37 (pp. 93-8)

1. *Memoirs of the Literary and Philosophical Society of Manchester,* Second series (1831), p. 384.
2. *The Manchester Chronicle,* 16 April 1831.
3. *The Manchester Times,* 16 April 1831.
4. *The Manchester Guardian,* 16 April 1831.
5. *Ibid.,* 16 April 1831.
6. *The Illustrated London News,* 27 April 1850.
7. *The Annual Register,* 1850.

CHAPTER 38 (pp. 98-101)

1. *Dictionary of National Biography.*
2. S. Plimsoll, *Our Seamen – An Appeal* (1872).
3. *The Illustrated London News,* 14 June 1873.
4. *Ibid.,* 26 April 1873.
5. *Ibid.,* 14 June 1873.
6. *Ibid.,* 24 July 1873.
7. *Ibid.,* 31 July 1873.

CHAPTER 39 (pp. 101-7)

1. G. W. C. Kaye, *X-Rays* (1923), pp. 24 ff. Also W. C. Röntgen, *On a New Form of Radiation, The Electrician,* 24 Jan. 1896 and 24 April 1896. Also O. Glasser, *Wilhelm Conrad Röntgen* (1933), pp. 11–48.
2. Editorial, *British Medical Journal,* 4 Sept. 1948.
3. Lord Rayleigh, *Proceedings of the Physics Society,* vol. XLVIIII.
4. S. P. Thompson, *The Saturday Review,* 11 Jan. 1896.
5. Lord Rayleigh, *The Life of J. J. Thomson* (1942), p. 65.
6. Editorial, *The Electrician,* 10 Jan. 1896.
7. *Punch,* 25 Jan. 1896.

CHAPTER 40 (pp. 107-12)

1. O. Lodge, 'Becquerel', in *The Chemical Society's Memorial Lectures 1908–13* (1914), p. 248.
2. R. J. Strutt, *The Becquerel Rays and the Properties of Radium* (1904).
3. Sir H. Dale, 'Accident and Opportunism', *The British Medical Journal,* 4 Sept. 1948, p. 453.

CHAPTER 41 (pp. 112-20)

Works consulted:
Statements Relating to the Atomic Bomb, H.M.S.O. (1946).

The Times, 7, 8 Aug. 1945.
Kessing's *Contemporary Archives*, 1943-6, pp. 6201, 7368.
O. Hahn, *New Atoms* (1950).

CHAPTER 42 (pp. 120-7)

1. H. Dircks, *The Life, etc., of the Second Marquis of Worcester* (1865), p. 579.
2. J. T. Desaguliers, *Experimental Philosophy* (1744), vol. II, p. 465.
3. *Ibid.*, p. 533.
4. S. Smiles, *Lives of the Engineers – Boulton and Watt* (1878), p. 79.
5. J. Muirhead, *The Life of James Watt* (1858), p. 21. Also F. Arago, *Distinguished Scientific Men* (1857), p. 521.
6. J. Cox, *Mechanics* (1904), p. 152.
7. A. Jamieson, *Elementary Manual on Steam and the Steam Engine* (1898), p. 154. Also W. H. Preece, *Watt and the Measurement of Power* (1897).

CHAPTER 43 (pp. 127-33)

1. E. A. Cooper, 'History of Cugnot's Engine', *Proceedings of the Institute of Mechanical Engineers* (1853), pp. 34-5.
2. F. Trevithick, *Life of Richard Trevithick (1872)*, vol. I, pp. 147-8.
3. S. Smiles, *Boulton and Watt*, p. 267.
4. F. Trevithick, *op. cit.*, p. 106.

CHAPTER 44 (pp. 133-7)

1. J. A. Paris, *Life of Sir H. Davy* (1831), vol. II, pp. 3-4.
2. J. Tyndall, *Faraday as a Discoverer* (1870), pp. 4-5.
3. J. H. Gladstone, *Michael Faraday* (1872), pp. 8-9.
4. H. Bence Jones, *Life and Letters of Faraday* (1870), vol. I, p. 49.
5. J. A. Paris, *op. cit.*, vol. II, p. 4.
6. S. Smiles, *Lives of the Engineers – Boulton and Watt*, pp. 198-9.
7. A. Carnegie, *James Watt* (N.D.), p. 93.

CHAPTER 45 (pp. 137-42)

1. H. C. Cameron, *Sir Joseph Banks* (1952), p. 146.
2. B. Franklin, *The Works of Benjamin Franklin* (1806), vol. III, p. 516.
3. H. C. Cameron, *op. cit.*, p. 51. Also J. Baron, *The Life of Edward Jenner* (1838), vol. II, p. 35.
4. J. A. Paris, *The Life of Sir H. Davy* (1831), vol. I, pp. 257 ff.

CHAPTER 46 (pp. 142-9)

1. Lord Macaulay, *History of England* (1889), vol. I, pp. 199-200.
2. E. Warburton, *Memoirs of Captain Rupert* (1839), vol. III, pp. 432 ff.

3. J. A. Paris, *op cit.*, vol. II, p. 23.
4. T. Thomson, *The History of Chemistry* (1831), vol. II, p. 144.
5. K. Jagow, *Letters of the Prince Consort 1831-61* (1938), p. 340.
Also T. Wemyss Reid, *Memoirs and Correspondence of Lyon Playfair* (1899), pp. 200-1. Also Charles Lowe, *Our Future King* (1898), p. 18.

CHAPTER 47 (pp. 149-53)

1. Zeller, 'The Paradoxes of Zeno of Elea', *Die Phil.*, vol. I, p. 540.
Aristotle, *Physics*, vol. VI, 9, p. 239.
2. Eutocius, *Commentary on Archimedes's Treatise on the Sphere and Cylinder*, ed. by J. L. Heiberg, vol. III, pp. 88-96.
3. Erastosthenes, *Theon of Smyrna*, ed. by E. Hiller, vol. II, pp. 3-12.

CHAPTER 48 (pp. 153-7)

1. *Hansard*, May 1866.
2. A. De Morgan, *A Budget of Paradoxes* (1872), p. 173.

CHAPTER 49 (pp. 158-61)

1. W. Hamilton, *Lectures on Metaphysics* (1870), vol. I, p. 169.
2. A. De Morgan, *op cit.*, p. 17.
3. Herman Lotze, *Logic*, trans. by Bosanquet (1888), vol. I, p. 307.
4. D. Stimpson, *Scientists and Amateurs* (1949), p. 53.